IN THE BELLY OF THE RIVER

Studies in Social Ecology and Environmental History

General Editors: MADHAV GADGIL and RAMACHANDRA GUHA

Studies in Social Ecology and Environmental History

IN THE BELLY OF THE RIVER

Tribal Conflicts over Development in the Narmada Valley

AMITA BAVISKAR

OXFORD
UNIVERSITY PRESS

OXFORD
UNIVERSITY PRESS

YMCA Library Building, Jai Singh Road, New Delhi 110001

Oxford University Press is a department of the University of Oxford. It furthers the
University's objective of excellence in research, scholarship, and education
by publishing worldwide in

Oxford   New York

Athens   Auckland   Bangkok   Bogota   Buenos Aires   Calcutta
Cape Town   Chennai   Dar es Salaam   Delhi   Florence   Hong Kong   Istanbul
Karachi   Kuala Lumpur   Madrid   Melbourne   Mexico City   Mumbai
Nairobi   Paris   Sao Paolo   Shanghai   Singapore   Taipei   Tokyo   Toronto   Warsaw

with associated companies in   Berlin   Ibadan

Oxford is a registered trade mark of Oxford University Press
in the UK and in certain other countries

Published in India
By Oxford University Press, New Delhi

© Oxford University Press 1995

The moral rights of the author have been asserted
Database right Oxford University Press (maker)
First published 1995
Oxford India Paperbacks 1997
Third impression 2000

ISBN 019 564 3925

Typeset by Rastrixi, New Delhi 110070
Printed in India at Pauls Press, New Delhi 110 020
Published by Manzar Khan, Oxford University Press
YMCA Library Building, Jai Singh Road, New Delhi 110 001

नदी का धर्म है कि वह बहती रहे।
काशीनाथ त्रिवेदी

It is the *dharma* of a river to keep flowing.
Kashinath Trivedi

Preface

Environmental movements in India assert that their ideology incorporates a thoroughgoing critique of environmentally destructive development. Such movements also claim that this critique is writ large in the actions of those marginalized by development — indigenous people who have, in the past, lived in harmony with nature, combining reverence for nature with the sustainable management of resources. Because of their cultural ties with nature, indigenous people are exemplary stewards of the land. I was moved by such beliefs when I started research in western India, examining the experience of a community of the Bhilala tribe which has organized to oppose the construction of a dam that threatens to displace them from their homeland. However, my expectation that I would encounter a community that lived in harmony with nature, worshipping it and using its resources sustainably, turned out to be both true and false. My neat theoretical framework linking nature–culture relationships to political critique, action and change crumbled into an untidy jumble of contradictions.

The dissonance between the depiction of tribal people in scholarly writing on the subject, and the everyday lives of tribal people, led me to ask the questions that run through the subsequent pages: Does the lived reality of tribal people today allow the formulation of a critique of development? What is the tribal relationship with nature today? How do people, whose struggles are the subject of theories of liberation and social change, perceive their own situation? When we place the present predicament of a tribal community in the context of its history of state domination, we see the contradictions inherent in the tribal relationship with nature — contradictions that permeate their consciousness as well as their practices. Given the problematic nature of tribal resource use, how accurately are the lives of tribal people represented by intellectuals in the environmental movement who speak 'on their behalf'? Only when we have a clearer understanding of the politics of representation can we forge an effective alliance between intellectuals and tribal people that is more just — to people and to nature.

Acknowledgements

The best part of writing a book is remembering everyone who made it possible. This work started as a thesis for Cornell University, which supported me generously through my graduate career. Cornell and the town of Ithaca were the best ivory towers that any scholar could desire. My mentor, Shelley Feldman, combined intellectual inspiration with warmth and friendliness. I am grateful for her support, especially the understanding with which she shared my dilemmas about fieldwork. I thank Philip McMichael for his insights into international political economy, and for encouraging me to see the world (system) in my grain of sand — Anjanvara. I also thank Mary Katzenstein for critically reading my work and taking a keen interest in it. The company of my fellow students at the Department of Rural Sociology, especially office-mate and buddy Jim Harkness, made work and leisure much more fun. Thanks to Dr G.G. Weix, kindest of friends and readers. I am grateful to Melanie Dupuis and Peter Vandergeest who gave editorial advice for parts of the first and final chapters. I thank Hugh Brody for letting me use parts of material originally written for him and the *Independent Review*.

I am grateful to the Institute for Intercultural Studies who funded my year of fieldwork in the Narmada valley. Being in the field was the best part of my life yet and I owe a debt of gratitude to the people who made it possible. I thank the Khedut Mazdoor Chetna Sangath for sponsoring my stay in Anjanvara. I am indebted to Rahul for all his kindness, for sharing with me his knowledge of Alirajpur. Special thanks to Chittaroopa Palit for thoughtful discussions and for demonstrating that praxis is possible. To Rakesh Dewan whose commentary and *kisse* let me appreciate the lighter side of alternative politics in Madhya Pradesh. To Amit for showing how culture can be creatively transformed for social action. I am deeply grateful to everybody in Anjanvara for their hospitality and help. To the children, especially Huma, Thumla, Lehria, Manya, Mohnia, Bachya, Guma and Nachria. To Dhedya, Bhangya and Budhya for their stories. And most of all, to Khajan and Binda and their family for welcoming me into their home and letting me stay month after month.

I thank the organizers of the Narmada Bachao Andolan for their help, without which my work would have been poor indeed. In Badwani, I am very grateful to Nandini Oza, Alok Agrawal and Shripad Dharmadhikary for their generosity and good company. In Kadmal, I thank Sitarambhai and Kalabhabhi and their family for their hospitality and their help. Thanks also to Devrambhai Kanera and Shakuntalabhabhi. Through the Sangath and Andolan, I have met truly heroic people who have blessed me with their friendship. Pre-eminent among them was Om Prakashji Rawal, whose untimely death cut short a lifetime dedicated to the socialist cause. Rawalji's vision soared as high as his ideals, yet he remained ever accessible, encouraging and inspiring everyone who had the privilege of knowing him.

Several people read the manuscript and suggested ways of improving it. Even though I sometimes disregarded what they said, their insights and expertise always encouraged me to write better. My intellectual debt to Ramachandra Guha is manifest in what follows. Andre Bétéille was ever generous with his time and attention in discussing problems along the way. Despite his busy schedule, Donald Attwood sent detailed comments which were of great help. Esha Bétéille's suggestions were invaluable for revising the manuscript. The book would be much poorer without Orijit Sen's illustrations.

Manasi and H.Y. Mohan Ram, passionate botanists both, identified the plants named in this book with consummate ease. As parents-in-law, their humour and indulgence created a truly congenial working environment. My parents, Kusum and Baburao Baviskar, kindled the flame of social inquiry and action in my mind. To two sets of parents and to Rahul Ram, with all my love, I dedicate this book.

Contents

List of Figures

Photographs by Rahul N. Ram
Illustrations by Orijit Sen

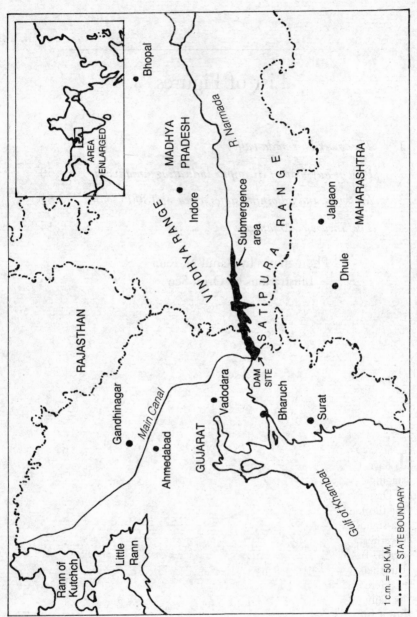

Fig. 1: The Lower Narmada Valley

Introduction

I have reached Alirajpur, the *tehsil* headquarters and an important staging post between the bountiful plains of Nimar and the trade centres of Dahod and Baroda. The bus-stand bustles as state express buses leave for the highway. The *paan* shop has a little crowd in front of it; the tea-stand boys rush with cups to the people who stop here so briefly. Overhead, flapping flocks of egrets whiten the tall *siris* tree where they perch. Below the tree stand private buses which leave from here too. But their destinations are more modest, as are their passengers.

Here is the bus to Bakhatgarh. There is still an hour before it begins its daily journey and yet it is already full, crowded with adivasis going home from the *haat* (market), or from business in the government offices and court. Bags bulge with chillies, tobacco, a mirror, jaggery, new shoes or some cloth, boiled sweets for the children. Ten minutes before the bus starts the driver swaggers in and yells at the conductor to hurry up.

The conductor has to sell tickets and pack as many people as he can into the bus.

Suddenly he becomes a foul-mouthed brute. He grabs an old man's head and shoves him forward, '*#@*#, are you deaf? Didn't you hear me telling you to move up? You *#@*# in the yellow blouse holding the brat, shut up and move!' Bullying, he jams people against each other as if they were cows to be herded — dumb, slow-moving animals understanding only blows and curses. Amazingly, the Bhil and Bhilala men and women who receive this treatment remain good-natured through it all, even giggling as they stumble across the aisle of the bus. The dignified look of the men with their bright turbans and moustaches, the strength of the women with their proud bodies, is at odds with their casual acceptance of this savage contempt of a *bazaaria* — a man who is not adivasi, of the town.

Meanwhile the bus driver condescends to talk to the other bazaarias — mainly clerks and forest guards who sit in the favoured row of seats in the front. It is a small world for them — these non-tribals posted in remote areas — for there are so few of their kind, and they are bonded by their superiority over the adivasis, for whose welfare they get their pay.

Every analytical enterprise is embedded in beliefs about the morality and the politics of research; every process of inquiry privileges a particular way of knowing the world. In the preface, I mentioned the theoretical concerns about development, environment, and indigenous resistance that prompted this study. Theories as well as the methodologies by which they are applied, authenticated and changed are all shaped by the ideology of the researcher. If we recognize that knowledge is socially constituted, historically situated, and informed by conflicting values, we are compelled to acknowledge that research cannot be the search and discovery of a single universal 'truth'. Instead, analysis has to be conceived as a process that mediates between at least two different, yet mutually conditioned, subjective views — those of the researcher and those of the people who are being studied. Here I address the issue of intersubjectivity by making explicit the connections between ideology and research methods, and by critically examining the relations between researcher and researched. In later chapters this analysis is continued in a discussion of the way in which the experience of fieldwork is represented through writing. Reflecting on the reflexivity of research sets the stage for the subsequent discussion.

Positivism *versus* Critical Inquiry

This inquiry emerges out of an epistemology that discards positivism in favour of more self-critical inquiry. The critical stance pertains to three related issues: the nature of the 'reality' that is being studied, the relationship between that reality and the researcher, and the ideological stance of the researcher. Discussing these issues reveals the process by which research is done, showing that behind the presentation of a seamless, authoritative account lies 'unruly experience transformed'. I ask, as James Clifford does, 'How, precisely, is a garrulous, overdetermined, cross cultural encounter shot through with power relations and personal cross purposes circumscribed as an adequate version of a more-or-less discrete 'other world', composed by an individual author?' (Clifford 1983: 120).

Unlike the positivist perspective, I believe that there is no unified tangible reality 'out there' for all to see, one reality which can be fragmented into independent variables and processes for the ultimate purpose of prediction and control. There are multiple realities, constructed by people in different ontological positions; inquiry into these multiple realities does not seek to discover a unified truth but is aimed at enriching our understanding of divergent, socially-situated truths (Lincoln and Guba 1985: 37–42). If the world consists of many contested realities, all of them backed by different groups of people, which version do we privilege? In the account that follows, I have mainly chosen to ignore the various ways in which the state regards adivasis, favouring a view that focuses on the ways in which adivasis regard the state. This 'bottom-up' perspective is seen as a corrective to the surfeit of accounts that uncritically accept the state's representations of adivasis.

Revealing these choices shows that, as researchers, we have to adopt a more self-conscious attitude towards the objects of inquiry, recognizing the dialectics of our relationship with them. Whereas positivism assumes that the unclouded gaze of the detached observer is best suited to capture 'objective' truth, this is merely a belief that has facilitated the domination of the learned 'outsider' as the expert authority. Truths that are contextual can only be known through intimate engagement with the perspectives of those whose lives are the objects of inquiry. However, this meeting of minds occurs on unequal terms. People's versions of reality are apprehended, interpreted and presented by the researcher who has her own preoccupations and presumptions. Also, that reality *includes* the researcher. We are not 'outside' the object of inquiry but are intrinsic to it; the joke goes that every indigenous community now consists of itself and a resident anthropologist. The world of the adivasis includes the researcher as well as the state; just as we have made 'the state' problematic, so must we think of the presence of the researcher as worthy of critical consideration.

While positivist scholars assume that inquiry can and ought to be value-free, I work with the explicit recognition that research is inherently a process guided by the values of the inquirer, values that manifest themselves in the formulation of the research agenda, and in the choice of theory and methodology. Belief in neutral and objective social research only obscures the unconscious biases of the inquirer.

Honking shrill departure, the bus trundles out of the bus-stand. We are finally off, the bus on the good metalled road, rolling through open spaces where toddy palms punctuate the distance, the wind blowing cool and dusty. I am travelling with two activists of the Sangath, on their way back to Attha from an Andolan meeting. I had written to them before we met, and they have agreed to introduce me into the villages.

Half an hour later, we are in the main street of Umrali, renowned for its Saturday cattle haat. It is quiet today, the police station and the Primary Health Centre look deserted. Out of Umrali the hills start — round, bare and dark brown. The bus climbs and we are in Sondwa, the Block headquarters, a smaller Umrali. Sondwa has the usual cluster of administrative buildings, and tea shacks for those who work in them. Overlooking them stands a crumbling feudal manor — the estate of the former *thakur* of Sondwa, once praised by the British for its mango orchards.

The feudal manor now rests decrepit upon the hill; the white-washed buildings of the modern state are emblazoned with hearty maxims – DRINK CLEAN WATER AND DEFEAT GUINEAWORM! And they are alive with activity below. Men wait on verandas, squatting patiently to see the *patvari* (revenue official) or the doctor. In this unfamiliar milieu, adivasis venture only when they must get something out of a hostile bureaucrat. Those who speak only Bhilali come with a *vaataad* in tow.

A vaataad (literally, 'talker') is an adivasi who speaks Hindi and who can confidently deal with clerks, peons and petty officials, buying them *bidis* and tea, cajoling them to accept an application, issue a receipt, look into a file. Only with supplication and the greasing of palms is business done.

Leaving Sondwa, we leave behind the metalled road, bouncing on a dirt track that is closed during the rains and after — what is marked on maps as a 'fair weather road'. The bus only runs on this route for part of the year. Now the driver changes to a lower gear; the engine whines in protest as it starts climbing. Outside the bus, the hills have grown paler and drier, the landscape lunar in its inhospitality. Rocks and sparse tussocks of grass — nothing else grows here. One hill is ridged with faint wrinkles — reminders of a failed forestry effort at controlling erosion and planting grasses to stabilize the soil. The raised scars of earth are all that remain of the attempt to recreate the forest that stood here not long ago.

Ideology and Inquiry

The concerns that guide the formulation of a research agenda can be traced to ideologies underlying academic political economy. George Marcus argues that issues in social science reflect a world increasingly pulled asunder by the expansive tendency of the global capitalist political economy to incorporate everyone into itself (Marcus 1990: 167). The world of larger systems and events no longer merely impinges upon and constrains 'little communities'; it is becoming integral to them. 'Little communities' — and subordinated people in general — are the besieged strongholds of autonomous cultural traditions. The research of critics of capitalism scales the bounds of beleaguered consciousness, for the quest for the native's point of view has now become a search for an authentic critical theory, embodied in the lives of those on the margins of capitalism. Thus, 'it is a seductive idea at the moment to liberal and radical cultural critics in search of some direction that the necessary insights are *there* in the lives of subjects, to be unearthed by careful interpretation . . .' (Marcus 1990: 185; emphasis in original).

Such an understanding tries to synthesize two divergent perspectives: one, the anthropological understanding of culture as autonomous, enduring over time, 'not without its own internal contradictions, but at least with its own integrity against the world'; and the other, the Marxist view of culture as a product of struggle. This merging, when successful, endows the research enterprise with a new legitimacy, for it can claim to make a localized critique known to the rest of the world.

My interest in the lives of adivasis, as shaped by development, and in their struggle through the Narmada Bachao Andolan — the movement against the dam that threatened to displace them, emerged out of such an interpretation of culture. In 1984, as a member of Kalpavriksh, a Delhi-based environmental action group, I participated in a month-long study of the Amarkantak plateau — the source of the river Narmada, and a Hindu pilgrimage centre in a forest inhabited by Gond and Baiga adivasis, now being mined for bauxite. This experience made me alive to the fact that the conflict over nature is not only over material resources but also involves fundamentally different cultural conceptions of nature. From this beginning I came to be interested in indigenous movements, which I identified as resistance to culturally destructive development. While studying in Delhi, and later in the United States, I kept abreast of events in the Narmada valley through Kalpavriksh. Nine months before I started fieldwork I visited the headquarters of the Andolan in Badwani, where I met and travelled with Medha Patkar, the leader of the movement. After discussing my still nebulous research agenda with her and other activists in Delhi, I wrote to activists of the Khedut Mazdoor Chetna Sangath in Alirajpur, who have organized the hill adivasis. My intention was to concentrate upon people who were threatened by the dam, who lived as communities of adivasis, not assimilated into the bottom of the Hindu mainstream.

Dusk falls as we reach Attha, the village that is the base of the Sangath. We pass some scattered houses, blurred now, and my companions call out greetings to the people inside. We cross a stream that leaves our feet slippery, then we climb up to a house on a hill, surrounded by a field of tall *tuvar*. We enter a room lit by the dim glow of two spirit lamps. There is a woman, lying down. A quiet man, dark and bearded, and another who is doing most of the talking, something about James Petras' article in *EPW (Economic and Political Weekly)* about the retreat of intellectuals from working-class politics. There is a picture of Bhagat Singh on the wall. Clothes hang from the rafters, including a pair of old jeans cut off at the knees. So far everything has been so strange that now my mind only alights on what seems familiar — the academic discussion, the pair of jeans, deriving a certain reassurance from the known.

The kitchen is a little different from the only adivasi home I have been to so far; it has the same wood-burning hearth for cooking, but there is much more stainless steel, bottles of home-made pickles — the latter most evocative of being away from home, of fond mothers somewhere in the background. There is a smaller room off the main room, filled with books. My companions recount the events of the Andolan meeting; I listen quietly, somewhat taken aback by the difference between my perceptions and theirs. While they talked, they have been busy cooking dal and wheat rotis. We eat. They roll out thin mats and I my sleeping bag, and go to sleep. I am enchanted by it all.

The next morning, we go to swim in a sunlit pond bordered by old mango trees, then bathe and wash clothes in a stream. Then, back at the house, someone asks me to give an account of myself, what I plan to do, and why. So far I have been ignored, my presence not directly acknowledged. This is the moment when I must be persuasive, must get them to accept me. Stumbling, struggling to formulate theoretical arguments in my sad Hindi, I try to explain my intention. Someone challenges me about my conception of hegemony and we argue for a while. I am more at ease then; after four years as a graduate student, this is familiar terrain. I encounter scepticism; they know me only as someone who wrote a letter from the USA. I am told that research is elitist, even exploitative, unless it is made subservient to practice. I can only concur, for I am conscious that they are doing something infinitely more worthwhile. Finally, one person says that they have no objection to my staying in a village, as long as I give something back to the people who are the objects of my study. It is decided that I should teach people to read and write.

As we wind up our discussion, adivasis start arriving for a meeting. After the activists have talked it over with other people, I am introduced to Khajan, the man from Anjanvara, the village which seems to be appropriate for my purpose. So it looks like I can start work right away. Soon the room is crowded. I sit in a corner and vainly try to follow the proceedings. I cannot understand Bhilali and I give in to a momentary wave of self pity, surrounded by strangers, feeling that my research is stupid and impossible.

Ideology and Methodology

Just as the ideology of the researcher is reflected in the choice of the subject of inquiry, and in its particular theoretical formulation, ideology is also mirrored in the methods by which research is conducted. As regards my study, giving priority to the conflict between the adivasis and the state necessitated that research be mainly conducted in the setting of the Narmada valley, for the meanings that people attribute to things and events can only be revealed in their social context. For the same reason, I chose inductive rather than deductive analysis because it allowed theory to emerge from the data in which it was grounded. Given the epistemological and theoretical values undergirding my research, it was apparent that the primary mode of data collection had to be participant observation, which I supplemented with informal discussions, the collection of oral histories and myths from villagers, backing up these with historical research at the National Archives of India.

The term 'participant observation' serves as shorthand for a continuous movement between the 'inside' and the 'outside' of events — on the one hand, grasping the sense of specific beliefs and practices empathetically, and on the other, stepping back to situate these meanings in wider contexts. As Clifford remarks, 'Understood literally, participant observation is a paradoxical, misleading formula. But it may be taken seriously if reformulated . . . as the dialectic of experience and interpretation' (Clifford 1983: 127). The emphasis that experience cannot be taken for granted, and that the researcher plays a crucial role as the academic interpreter of culture, suffuses the self-conscious ethnographic writing current today.

The choice of ethnographic methods led me to concentrate upon the intensive study of a village community, an analytical unit that I gradually came to perceive not as an entity that could be taken for granted but as a lived and contested reality. However, I could not be content with simply presenting another, albeit more complex, view of the community; the aim of the inquiry could not merely be interpretation *ad infinitum*, there had to be a movement from understanding to action. Recognizing the politics of scholarly research, I felt that involvement with the work of the Sangath and the Andolan would be a practical antidote to academia's tendency towards quietism. At its best, besides articulating people's lived concerns, fears and aspirations, critical research could act as a catalyst for ideological reflection and action (Friere 1970). Believing that research could be part of praxis in a process of empowerment and transformation was important, for I felt the opposite only too often. I had the luxury of indulging in intellectual pursuits, I was free to read and reflect and observe without giving anything back. And I had the freedom of being able to leave, of going back to the city to be admired by relatives and friends for living in the forest — a place without bathrooms — the ultimate middle-class nightmare. My doubts as well as my beliefs led me to seek the support of activists directly engaged in political mobilization among the adivasis, and against the dam.

We set off the next morning, Khajan, an activist and I. I was told that it took five hours to reach Anjanvara, something that brought home to me the difficulties of organizing here where so much effort is spent in simply going from one place to another. However, my presence made it a much longer affair. We had to stop frequently so that I could catch my breath and rest my knapsack, and everybody had to slow their pace to let me keep up. Later, when I was on my own, I took six hours over the same path — stopping by streams and on top of slopes, looking at the changing views before me. We climbed steeply and descended through two villages, then there was a long fairly level stretch after which we went downhill quite rapidly, crossing many streams and passing through some fairly dense forest. Finally, after negotiating all that, we suddenly reached the river — a sight so beautiful that my spirits soared. A wide valley with a sandy stretch littered with rocks through which flows the river — big, peaceful and empty. Across the river are tall hills, dotted with the green of teak and *anjan*, a green that is echoed in the river. We bathe in the river, then set off again towards Anjanvara.

The Power of the Researcher

The resolve that this research be oriented towards praxis was easier made than done. During the research process there was a constant tension between my initial theoretical assumptions about development and resistance, and my simultaneous resolve to imbibe 'native' theories. This tension clouded the very validity of interpretation: when our understanding diverged, whose dominated? My belief in multiple realities was checked by a growing realization that a plurality of voices did not always come together in harmony. A dialogical negotiation of meaning did not necessarily lead to an easy resolution. As my presuppositions were challenged by the experience of fieldwork, they had to be modified — a process that resulted in reflection and theorizing about the relationship between the ideological assumptions of researchers and how they represent the people whom they study.

My first week in Anjanvara. I have been thrown into the deep end. Only Khajan speaks Hindi and I have to communicate with others through him as I try desperately to learn Bhilali. We were sitting outside one night; Khajan was describing an aeroplane. He has been to Bombay for an Andolan demonstration. Someone asked how many people could sit in- side one. I answered. Someone asked what the fare was. I guessed, maybe four thousand rupees to go from Delhi to Bombay. There was a wondering silence. Then someone said, it must have more than one engine otherwise if the engine failed it would crash. Someone asked how much a pilot earned. I said I didn't know, maybe five thousand rupees a month. Again a silence. So I felt constrained to qualify that by pointing out that, in Bombay, you spend a lot of that money buying things that people grow or make for themselves in the village. A lame excuse, and definitely prompted by guilt. I find myself lying all the time. People keep asking me what my things cost — my sleeping bag, the bangle that I wear, the soap that I use. I can't bring myself to tell the truth, so I always scale prices down. They are still shocked. I am embarrassed by my riches.

My possessions contrasted sharply with the material poverty of the people with whom I lived. In their universe I was powerless and unskilled — my attempts at turning the grindstone would send women into peals of laughter. But despite my lack of skills, I could come and live with them and go away. They had no reciprocal power to enter my world of the university. This was the fundamental inequality: that the social arrangements of class determined that I had mobility and freedom *because* they did not. When the exploitative and intrusive aspect of fieldwork seems inescapable, how does the researcher deal with the in- equality of this relationship?

Acknowledging the hierarchical and power-laden relations be- tween researcher and researched, Clifford draws attention to 'the historical predicament of ethnography, the fact that it is always caught up in the invention, not the representation, of cultures' (Clifford 1990: 2). According to Clifford, the solution to this dilemma is to fully accept the limitations of the research process and product and to reduce their claims, while perceiving subjects as collaborators in the project. Thus there has been a move in ethnographic writing from the expository conventions of 'objective' author-evacuated texts to author-saturated 'dialogical' ones (Geertz 1988: 9). But a sensitivity to power inequalities serves to undermine scholarly pretensions about alliance and collaboration with the Other (Stacey 1988: 25). We only acknowledge the ethical dilemmas of research, we do not resolve them, for in the text the voices of the people are ultimately translated by the researcher. Finally, it is the author who prevails.

However, recognizing the limitations of the research process and product is not to surrender the overall validity of the scholarly enterprise. An author may have the power to decide how people should be represented, but the researcher's partial truths cannot be lies, for the world does not tolerate all representations equally. The power of the researcher over the text can never be absolute. All texts remain vulnerable to being disputed and discredited. In addition, we must remind ourselves that the scholarly text is simply one representation among many. To think that it is fatally flawed by its shortcomings is to overrate its importance. One book is far from being the only, definitive, representation of people; they act as subjects of their own accord and speak for themselves, and they will go on doing so. They will also continually be represented by others. While we should be sensitive to the contradictions inherent in the researcher/researched relationship, we cannot allow our ethical dilemmas to immobilize us.

Research and Reciprocity

Being in Anjanvara, and in the Narmada valley, was crucial to my research. What did I do in return for the people whose generosity made this possible? My most satisfying and most frustrating experience in Anjanvara was to teach people to read and write. When I came to the village, only one boy, who had lived for a while in another village which had a school, was partly literate. On my very first night, men assembled for a meeting. Khajan explained that I wanted to see how they lived and that I would teach during the nine months that I would stay in the village. Within an hour, people had decided to collect money from every household to buy slates and chalk. Two young men were pressganged into going to the haat the very next day to buy material for the entire village. Before the week was out, we were ready to start.

Two weeks and I am already sick of it. Twenty children come every morning — including two girls. All of them are of different ages and aptitude. When I start explaining things to one, others get bored. To make things easier, I am teaching them to read and write Bhilali, with which they are familiar, using the Devnagri script. I instruct them in my broken Bhilali liberally interspersed with Hindi and somehow we communicate. Almost the very first sentence that I learn in Bhilali: 'Will you stop fooling around and pay attention?' In the afternoon come the young men, about six of them. I go over the alphabet again, trying to establish the connection between letter and sound. Slowly, painfully, they copy out different letters, learn to write their names. Then, in the evening, Khajan, four other men, and three girls come to study. I repeat the entire exercise with each one. The girls come one day and stay behind at home to work the next; the irregularity means that each time we have to start anew. We peer at the slates by the glow of a spirit lamp. I am exhausted. When I go to any woman's house and ask her to join the evening class, she says that she would like to learn but only if I could come to her house and teach her there in her spare time. This is enthusiasm indeed; how I wish there was less of it.

A truly ghastly session today! I thought I would inject some fun into the tedium of learning by getting the kids to draw and sing. Well, the drawing part was fine. It was quite a thrill to see what they drew for the very first time in their lives, without any intervention on my part — minimalist stick figures of dogs, people, goats and trees. Then I asked them to sing, using the Hindi (and I thought also Bhilali) word for song — *gaana*. They kept protesting but I thought they were just being shy, so I kept insisting. Then they started singing the creation myth that I had heard on my first day [see Chapter 7], breaking off every now and then to say, 'It's coming, it's coming'. I didn't understand and I urged them to go on. Then, suddenly, Lehria got possessed! One moment he was singing, the next he was hopping up and down barking like a dog, his body shuddering violently. Chaos. Someone's slate went flying into the air, a pot of water was knocked over. Meanwhile, a few of the boys were trying to appease the God possessing Lehria, throwing water and bowing before him, whacking him on his back. Just as suddenly as it started, it stopped. Lehria's body went limp. Everyone turned on me: 'You asked us to sing and summoned God idly! Look what happened!' Then Khajan came home and I appealed to him. He explained that I should have asked them to sing *geet* (a Hindi and Bhilali word for song); while gaana simply means 'song' in Hindi, it has a somewhat different meaning in Bhilali, that of *gayana*, the song of creation sung as part of religious rituals, which induces possession. I was mortified. To make things worse, Jevanti went around recounting this to everybody, telling them with great glee, 'Amita was really scared!' 'Of course, I was not!', I replied crossly.

The evening sessions are suspended! The Gulf crisis has made kerosene scarce from the haat, so there is no spirit lamp to teach by. I can hardly hide my relief. I seem to do nothing but teach all day. I didn't come here to do this. When one of the activists visits the village, he tells me to stop taking this so seriously. But I can't turn people away. I reflect on what I am trying to accomplish. Romka has told me that she spares two of her sons from work because she wants them to become literate, go out of the village, and get government jobs. Apart from the fact that nine months of learning is not enough to qualify Romka's sons for government jobs, is this what literacy should be oriented towards? Adivasis who are educated in schools usually become estranged from their culture; as petty officials they exploit their own people. Besides teaching skills, education must reinforce pride and confidence in people's knowledge, and foster an adivasi political consciousness. It must be made relevant to people's lives.

Teaching in Bhilali helps me to learn the language more quickly. Very soon, the teenage boys understand the logic of spelling. They make lists — of trees, of people in their home, of things that they buy from the haat, of villages. I dictate sentences and they write them down. Soon, they are good enough to write on paper with pencils. Now they have a record of what they have learnt. On Sunday, we play games — *kabaddi*, *kho-kho*. They bring bows and arrows and have archery competitions. On my first trip out of the village I buy paints and chart paper, and we slowly start drawing and painting a large map of the village. They carefully draw each house and write the name of the family below. They draw and name all the trees, show people working in the fields, fetching wood from the forest, water from the river, dancing. As it slowly takes shape, I can see that it is going to be magnificent. Every morning when they leave, we call out Sangath and Andolan slogans, and the valley rings with their voices.

In the end, my teaching ended up concentrating on the children who came regularly in the mornings. I would feel gratified when they accompanied adults to political meetings outside the village, their faces looking tiny and unfamiliar underneath the massive turbans that they wore on formal occasions. The children stayed for much of the Sangharsh Yatra, the month-long rally organized by the Andolan. However, when they came back from the Yatra and I quizzed them about their impressions, they would only sing garbled versions of Andolan songs and do ribald imitations of some of the principal speakers. I ruefully concluded that this too was political education! Near the end of my stay, the boys of Anjanvara attended the *jangal mela* — the annual meeting of the Sangath where people from all the member villages come together to enjoy themselves and to talk about political matters. No other village brought young boys along. The children from Anjanvara participated in the archery contests, where they covered themselves with glory by competing against each other and walking off with all the prizes — fruit-tree saplings.

Despite its ups and downs, despite its limited scope, teaching did help people a bit. It demystified the world of letters and introduced people to the power of being able to decipher tea-packet labels, posters, bus destinations, and put a signature instead of a thumb impression. It made parents proud; every now and then someone would stop by and take in with immense satisfaction the spectacle of their offspring writing industriously. When I went to neighbouring villages, people would try to lure me away from Anjanvara, pointing out the superior qualities of their village and their children. I think that, because of my role as teacher, my presence in the village was not felt to be oppressive.

Tensions between Participation and Observation

I had begun discussing the issue of methodology by stating that the best strategy for observing the politics of the Sangath and the Andolan was participation in their activities. However, my position as a researcher who was involved in the lives of people in the valley primarily out of scholarly interest, someone who would leave after a while to write critically about them, put a constraint on some of our friendships. When Sangath activists had been arrested [see Chapter 9], I had postponed going to Anjanvara so that I could assist them. After their release, I was impatient to return to the village and resume my work. But some of the activists felt that I should stay on to help out. My reluctance must have made itself known; someone said that my role as a researcher, 'always sitting there with notebook and pen', created a distance between myself and the Sangath. I acknowledged the truth of this even as I felt resentful about it.

During the process of writing, I have felt anew the painful contradictions of my position. The greater the intimacy, the greater the danger of betrayal. I was treated with warmth, openness and generosity by my friends in the Andolan and the Sangath. How could I pay them back with criticism? Who would trust me again? Who should I be true to — my friends or some obscure intellectual ambition? My privileged position as an academic allows me the luxury of critical evaluation; they are defending their life's work.

Finally, the exigencies of participation have impressed upon me the necessity of moving from interpretation to action. I had earlier chafed at the theoretical compulsion to apply closure, to render life intelligible within a text, imposing 'order through writing on a world whose essence is its fragmentary character' (Marcus 1990: 191). Yet engagement with the work of the Sangath and the Andolan showed me that open-ended interpretation was not enough. Theoretical concerns had to be grounded, their implications had to be relevant to people's lives. After reading the account in the following pages, people should not ask 'So what?' The answer must be manifest.

Outline of the Discussion

After a brief look at the Indian experience of development and its impact on poverty and the environment, I shall examine the ways in which this experience has been understood and explained by Ecological Marxists. This theoretical approach outlined in the next chapter will inform the discussion that follows. The discussion in Chapter 3 begins by tracing the history of adivasis in the submergence area of Sardar Sarovar Project, delineating their relationship with changing states, and the gradual alienation of their natural resource base and its increased incorporation into a market economy. It examines the subjugation of adivasis to the increasing extractive and executive powers of non-local, bureaucratic structures which predate colonialism. This history looks at the ways in which the expansion of alien state power, which marginalized adivasis, was continually resisted — seeking in that rebellious tradition a resource facilitating collective action in the present. The adivasi experience of marginalization and resistance has resulted in the crystallization of an adivasi community and consciousness, distinct for its opposition to the mainstream yet shaped by the fact of its domination. The next chapter discusses this contradictory relationship between Bhilala adivasis and the Hindu mainstream, exploring the ways in which Bhilala life in the hills today is the basis of a separate identity despite processes of Hinduization.

The next three chapters on community, economy and ideology, depict the present cultural being of Anjanvara, the Bhilala village where I stayed in the hills of Alirajpur. The life of the village represents the adivasi history of political and ecological marginalization, for Anjanvara lies in a forest degraded by the state, and people cultivate land and livestock on hill slopes with thin soils. Yet, despite the poverty of their resource base, people have created a production system remarkable in its diversity and self-sufficiency, utilizing different aspects of the land, forest and river to the fullest. The chapter on community studies how such resource use is enabled by co-operation structured around the patriarchal clan, a kin group bound by norms of reciprocity and mutual aid. However, the spirit of generosity towards one's kin contains within it calculation and conflict. Households in the village manoeuvre to improve their standing *vis-à-vis* each other, through marriage alliances and the accumulation of goodwill. As women are central to the continued survival of the community, much of the local politics revolves around the defence of honour, partly defined in terms of control over women. Besides mobilizing everyone within the adivasi community, the politics of honour allows women some degree of freedom from the structures of patriarchy. As community is defined in terms of the corporate unity of the patriarchal clan, by manipulating male honour, women's actions both enable and challenge the system.

The dependence of the village community on the natural base which sustains its life is acknowledged through religious beliefs, and rituals that seek to secure nature's co-operation. Yet rituals of reverence prove inadequate for managing natural resources. The chapter on economy and ecology describes how the state's previous destruction has so exhausted the forest that even the modest demands of everyday use by adivasis deplete natural resources beyond repair. The alienation of the forest by the state has forcibly confined adivasi cultivation to fragile hill slopes, and the slow decline of the relatively self-sufficient hill economy has drawn people against their will into the ever-widening circles of com-modification of their produce as well as labour. This chapter examines how the continuity of the present way of adivasi life is ruptured by state-induced processes of ecological deterioration.

The chapter on nature and ideology elaborates the contradic-tions between practice and beliefs about nature. While people revere nature in all its forms, and religiosity suffuses their everyday lives, these beliefs do not translate into a set of sustainable resource use practices. People acknowledge the power of nature, whose uncertainties rule their lives, and they make strenuous efforts to seek its co-operation. Yet, ritual propitiation is insufficient for dealing with the present predicament of resource degradation. By depicting the disjuncture between adivasi theories of nature, and their inability to derive a sustainable livelihood, this chapter shows how the reality of adivasi life contrasts with environmentalists' claims about it.

If the chapters on community, economy and ideology are about *being*, then the following chapters are about processes of *becom-ing* — the transformation of adivasi life through political collec-tive action. In the earlier discussion of community life we had seen how political conflict is organized around honour. Even though it did not seem to fit into the theoretical structure of research, it became necessary to include the politics of honour in the present analysis because it overwhelmed the experience of fieldwork. Honour could not be ignored since it was intrinsic to the way in which adivasis acted. In fact, to excise honour from adivasi politics would be to collaborate in its misrepresentation. The chapter on Sangath politics describes the ways in which the consciousness of the adivasi community is transformed, from the pursuit of honour to action against the state, fighting to secure access to the land and the forest. This chapter examines the relationship between the adivasis and the activists who have facilitated their organization, and critically evaluates the Sangath's experience of collective action trying to resist the state and, at the same time, make it more accountable to the adivasi community. The chapter on Andolan politics examines the way in which adivasis have become a part of a larger, more diverse, movement against the Sardar Sarovar Project. The anti-dam movement has raised the broader questions about development and its critique with which we started this discussion. This chapter depicts the complexity of the issues which unite adivasis with other political constituencies, situated in different social spaces, and spread across the world, showing how the different ideological streams of environmentalism come together in practice.

The concluding chapter draws the themes of environmental practice and adivasi consciousness into the general theory of development and resistance. This final discussion analyses contradictions in the way in which the lives of adivasis are represented by intellectuals who speak 'on their behalf', and the problems and possibilities engendered in the process of coming together to formulate a critique of development. Finally, I end the discussion by proposing a more nuanced perspective on adivasi environmentalism which respects people's understanding of what they are fighting for, a perspective that builds upon their strengths while remaining conscious of their vulnerabilities.

2

National Development, Poverty and the Environment

I: Development, Poverty, Environment

In 1947, after years of bitter struggle, India finally became independent. In a voice charged with emotion, Jawaharlal Nehru proudly proclaimed to the nation that India had kept her 'tryst with destiny'. For the leaders of the nationalist movement as well as for the general populace, it must have been an exhilarating moment to stand on the threshold of a newly independent country and imagine all the possibilities for progress that lay ahead.

Now, forty-five years later, what have we achieved as a nation?

This chapter will discuss the choices made by the independent Indian state to develop the nation and improve the lives of its citizens. In the first part, I will examine the impact of development policies on the lives of the poor and on the environment, and the conflicts that have arisen as a consequence. In the second I analyse these conflicts using the theoretical approach of Ecological Marxism. This approach focuses on conflict over resources and gives primacy to popular grassroots struggles, the collective resistance of tribal communities, fisherfolk, labourers and peasants to processes which impoverish them and destroy their natural environment. I end by looking at the ways in which the lives of tribal people have been depicted in these theories as exemplars of an alternative paradigm of development based on harmony between nature and culture.

From its inception, the Indian state was confronted by two different visions of reconstruction: the Gandhian project of reviving the village economy as the basis of development, and the Nehruvian plan for prosperity through rapid industrialization. On 5 October 1945, Gandhi wrote a letter to Nehru in which he outlined his dream of free India:

I believe that, if India is to attain true freedom, and through India the world as well, then sooner or later we will have to live in villages — in huts, not in palaces. A few billion people can never live happily and peaceably in cities and palaces ... My villages exist today in my imagination ... The villager in this imagined village will not be apathetic ... He will not lead his life like an animal in a squalid dark room. Men and women will live freely and be prepared to face the whole world. The village will not know cholera, plague or smallpox. No one will live indolently, nor luxuriously. After all this, I can think of many things which will have to be produced on a large scale. Maybe there will be railways, so also post and telegraph. What it will have and what it will not, I do not know. Nor do I care. If I can maintain the essence, the rest will mean free facility to come and settle. And if I leave the essence, I leave everything (Chandra 1987).

'God forbid that India should ever take to industrialism in the manner of the West', Gandhi observed. 'The economic imperialism of a single tiny island kingdom (England) is today keeping the world in chains. If an entire nation of 300 million took to similar economic exploitation, it would strip the world bare like locusts' (Gandhi 1951: 31). The appeal of the Mahatma lay in his programme of revitalizing village communities and craft production by employing simple technologies to provide jobs and a decent livelihood to a predominantly rural population. The liberation that Gandhi promised was not merely an economic independence; it was, most profoundly, an assurance that the cultural traditions of the Indian peasantry would reign ascendant.[1]

Gadgil and Guha remark on the apparent paradox that the 'Gandhian era of Indian politics saw the juxtaposition of a peasant-based politics with the increasing influence of Indian capitalists over the Congress organization' (Gadgil and Guha 1992: 182). Despite the theoretical primacy of the peasant, it was the Indian industrial class that was able to use nationalism to wrest concessions from the British. The protective measures reluctantly imposed by the British on the post-World War I economy enabled Indian manufacturers to amass respectable fortunes and strengthen their political hold. The expansion of the Congress was financially assisted by Indian capitalists whose initial distrust of the party dissolved when they realized that Gandhi's politics consisted of an all-embracing effort to mobilize all sections of society in a common struggle. Gandhi's theory of the capitalist and the landlord as 'trustees' of national property was ambiguous enough to accommodate potentially divergent interests.

Despite the deification of Mahatma Gandhi in the pantheon of nationalism, his vision was eclipsed within the Congress party by the ideas of Nehru, Sardar Patel and others. Gandhi's vision struck no chords in the mind of Jawaharlal Nehru, who replied rather brusquely to Gandhi's letter of October 1945: 'It is many years since I read *Hind Swaraj* and I have only a vague picture in my mind. But even when I read it twenty or more years ago it seemed to me completely unreal ... A village, normally speaking, is backward intellectually and culturally and no progress can be made from a backward environment' (Chandra 1987). Most Indian nationalists believed that India's reconstruction could only come about through an emulation of the West, 'intellectually through the infusion of modern science, and materially through

---

[1] Gandhi's 'message' was variously interpreted and acted upon, sometimes in ways unanticipated and unapproved of by the Congress. Amin (1988) discusses diverse religious and militantly political responses to Gandhi by the peasants of Gorakhpur.

the adoption of large-scale industrialization' (Gadgil and Guha 1992: 183). Through intensive industrialization and urbanization, fostered by a strong nation state, India could overcome the handicap of its colonial past to catch up with the West. The rapid strides taken by Germany under Bismarck, by Meiji Japan, and Stalinist Russia proved that this economic miracle was possible. Not surprisingly, this programme was enthusiastically supported by Indian capitalists, who foresaw that the Indian state's investment in essential infrastructure would encourage the flowering of private industry.

Gadgil and Guha describe the choices that were made at this crucial juncture by the Congress leadership. Its most important decision was the adoption of the 'industrialize or perish' model of economic development in the Second Five Year Plan. From the Second Plan onwards, the Indian government spent only 22 per cent of its total Plan budget on agriculture, even though 75 per cent of the population was engaged in agriculture. The greater share of Plan outlay consistently went to develop industries, which employed merely 11 per cent of the population (Kohli 1987: 73). This strategy indiscriminately applied 'modern' technologies, with little regard for their social or ecological consequences.

In theory there were, of course, many options available to the Indian state. The technologies adopted could be capital or labour intensive; they could be oriented towards satisfying the demand for luxury goods or fulfilling the basic needs of the masses; they could degrade the environment or be non-polluting; they could use energy intensively or sparingly; and they could use the country's endowment of natural resources in a sustainable fashion or liquidate them; and so on (Gadgil and Guha 1992: 184).

However, these choices were critically affected by three powerful interest groups: capitalist merchants and industrialists, the technical and administrative bureaucracy, and rich farmers. The shape of the Indian economy today is a direct consequence of the political choices made forty years ago.

Achievements after Independence

Forty-five years after independence, India is transformed in numerous ways. Mass starvation on the scale of the Great Bengal Famine

of 1942 is now a distant memory. This is partly due to a tremendous increase in foodgrain production since the 1960s. India's dependency on food imports has shrunk to such an extent that today
it imports less than two per cent of the food that it eats (UNDP
1993: 161). Virtual self-sufficiency in food production is matched
by a diversified industrial base. Over the last ten years, our Gross
National Product has been growing annually at the respectable rate
of 5.4 per cent. The improvement in the quality of most lives is
also reflected in health statistics: an average Indian, who in 1960
would have died at the age of forty-four, can today expect to live
for almost sixty years (UNDP 1993: 143). In many urban centres,
amenities like electricity, piped drinking water, public education,
health care and transport can be taken for granted.

For some Indians, the post-independence years have also
brought vast opportunities for enhanced consumption beyond the
basics. The middle classes and above can afford or aspire to possess
many more motorized vehicles, television sets, coolers, refri-
gerators and other durable goods. Glittering enclaves in cities like
Delhi or Bombay boast shops and lifestyles that rival New York
or Tokyo. Affluence unimagined in colonial times, except by a
handful of princely rulers, is the happy lot of India's elite.

All this can be attributed to the early emphasis on industrial
growth, both in manufacturing and in agriculture.[2] It was generally assumed that the benefits of industrial expansion, increased
production, employment and income, would 'trickle down'
through the economy to those at the very bottom. At the same
time, there was some attempt to temper the differences in a grossly
unequal society by undertaking a modicum of income redistribu-
tion. This was done through half-hearted land reforms, with the
Congress walking the tightrope between populist appeal and the
support of the landed classes. The zamindars, loyal to the former
colonial powers and therefore expendable, were deprived of land
above a prescribed ceiling, but the holdings of middle and rich
peasants were left intact. The status of most of the landless poor
remained unchanged.[3]

[2] The development of agriculture in India has been marked by the increasing
use of industrially produced inputs such as fertilizers, pesticides and heavy
machinery. The transformation of agricultural processes due to the application
of capital-intensive technology may be called the industrialization of agriculture.

[3] For a detailed evaluation of land reforms in three Indian states and their

For the most part it was believed that industrialization and urbanization were essential strategies for national development; their benign character was not questioned even though these processes were marred by undesirable 'side-effects' such as increasing levels of urban congestion, and air, water and noise pollution. Pollution was not perceived as a priority in the early decades of industrialization. While planners shrugged their shoulders and said 'one cannot make an omelette without breaking eggs', economists simply called these problems 'externalities' and banished them outside the realm of the rational.

Nevertheless, environmental pollution, lamented from the very first days of black-belching smokestacks in the Industrial Revolution, has become an issue on the international agenda, a subject of treaties and negotiations. Global warming threatens to bring about wildly fluctuating climatic conditions and the gradual submersion of Bangladesh into the sea. Ozone depletion has already sharply increased rates of skin cancer and cataracts, especially in the southern hemisphere (Gore 1992: 85). These problems, together with the clouds of acid rain that drift in defiance of international borders, have driven home to the West that environmental concerns need to be universalized. The futures of North and South are inextricably linked.

Developing countries such as India have viewed the industrialized North's environmentalism with suspicion.[4] Prescriptions for controlling air pollution entail curbs on the use of chlorofluorocarbons (CFCs), a class of chemicals used in refrigeration, and reductions in the emission of acid rain-causing sulphur and nitrogen oxides by installing costly scrubbers and precipitators. Defenders of Indian industry claim that these measures are uneconomical; they have chosen to see the issue as a trade-off between imperative growth and expensive pollution control. Indian delegates at the 1992 Earth Summit in Rio de Janeiro reacted as if a concern about pollution was a neo-imperialist plot, designed to keep the South ever backward. In this view, since pollution is a necessary concomitant of industrial growth, the North's attempts to reduce

---

impact on poverty, see Kohli (1987).

[4] India has been justly indignant about the double standards employed by the West. The industrialized nations have chosen to preach to the South about reducing methane emissions caused by paddy cultivation and bovine digestion, while ignoring their greater culpability in the production of carbon dioxide.

emissions globally would deny the South the fruits of development. Environmentalism is perceived as a luxury that the South, with its pressing problems of hunger and deprivation, cannot afford. This attitude stretches back in time to the 1972 Stockholm Conference on Environment when the Indian prime minister, Indira Gandhi, proclaimed that 'poverty is the worst polluter'.

The Persistence of Poverty

The optimistic assumption that increased welfare due to industrial growth would automatically percolate to the poor has not been borne out by experience. Both poverty *and* pollution continue to mar the Indian landscape. For the 423 million people below the poverty line, almost *half the nation*, development has been a distant phenomenon, watched from the wayside (UNDP 1993: 141). Even after forty-five years of independence, most of these people are trapped in a web of poverty and powerlessness, their faculties and talents buried under the crushing weight of everyday life. The profile of this deprivation has a numbing quality because of the sheer vastness of the problem: every year 3.84 million children die before they reach five years of age, killed by hunger and disease, another 73.1 million children under five are malnourished; 72.9 million children are not enrolled in schools; an even larger number do not go to school; 281 million adults, 61 per cent of them women, cannot read or write (UNDP 1992: 141). And India's boast of being self-sufficient in food is a cruel hoax for the hundreds of millions who habitually go hungry. While isolated reports of starvation deaths, those in Kalahandi and Palamau for instance, jolt the news-reading classes out of their complacency, the pervasiveness of poverty goes unnoticed most of the time.

Another hope belied is the expectation that economic growth would create jobs and income for the masses. While the South Asian economy has almost doubled in size between 1975 and 1990, employment has increased only by 37 per cent. This phenomenon has been called 'jobless growth' by the United Nations Development Programme (UNDP), which attributes it to the application of inappropriate capital-intensive technology. The UNDP *Human Development Report 1993* remarks that 'the prevalent technology reflects the existing pattern of income distribution — 20 per cent of the world's population has 83 per cent of the world's income and, hence, five times the purchasing power of the poorer 80 per cent of humankind. Clearly, technology will cater to the preferences of the richer members of the international society' (UNDP 1993: 37).

It is frequently argued that the benefits of economic growth have been neutralized by the high rates of growth of the Indian population. According to this view, poverty exists because of the pressure of ever-increasing numbers on a finite resource base. While population growth rates *are* a source of concern, they do not constitute a problem by themselves. Statistics of present food production and projections for the future show that India is more than capable of feeding its citizens adequately (UNDP 1993: 161). Yet the fact that this food does not reach the hungry points to another problem — social inequality. It is not so much population pressure that causes hunger but the *distribution* of food and the social and material resources needed to obtain it (Webster 1984: 102).[5]

The exacerbation of social inequality has been intrinsic to the process of development as experienced so far. Through centralized planning and execution, the Indian state retains control of the 'commanding heights' of the economy. Its projects and policies are largely devised by bureaucrats and engineers, usually in collaboration with big business and large farmers' lobbies, and with very little popular participation. The 'industrialize or perish' model has been combined with a tendency to be grandiose and build on a gigantic scale. The concentration of public resources in particular sectors, such as heavy industrial infrastructure and irrigated agriculture, has created projects such as the superthermal power complex in Singrauli and the Narmada Valley Project, which have a gargantuan appetite for public funds. One dam of the Narmada Project, Sardar Sarovar, will alone cost more than

---

[5] There is also an impressive body of evidence that shows that high rates of growth of population are a *consequence* of human deprivation, and that poverty alleviation directly affects population growth. In a study of West Bengal and Kerala, Moni Nag (1984) demonstrates that Kerala's success in population control is due to the state's investment in improving education and health care, especially for women. Kerala also ensures that food will be available and affordable to all its citizens through a public distribution system that covers 97 per cent of its population. While Kerala is much less industrialized than West Bengal, its people marry later and have fewer children.

the entire amount spent on irrigation by the government since Independence! Despite evidence to the contrary, the state still prefers to believe that showy, expensive projects are better investment than more modest, decentralized ones. For instance, in a distribution of resources that defies all logic, during 1951–85 the Indian government invested the greater part (64 per cent) of its total irrigation outlay of Rs 23,180 crores on major and medium irrigation projects which have irrigated land at the cost of Rs 19,310 per hectare, even though minor irrigation projects provide water at a much more economical rate of Rs 4520 per hectare. Despite the greater outlay of 64 per cent, major and medium irrigation projects have *irrigated only 30.5 million hectares of land, while minor projects have irrigated 37.4 million hectares*! (Sachidanandan 1988: 80).

The skewed priorities of the government are also reflected in the pattern of investment in energy production. Only three per cent of India's energy needs are met by electricity, while biomass provides more than 50 per cent. Yet, in the Seventh Plan (1985–

1990), Rs 32,000 crores were allocated for the electricity sector, whereas the development of biomass resources received less than Rs 2000 crores (Dharmadhikary 1991: 15). Most of this power is supplied at highly subsidized rates to urban consumers and industries, and for agricultural pumping. Of course, the poor who depend primarily on biomass for their energy needs, receive no subsidy at all.

Poverty and the Ecological Crisis

This pattern of development has fundamentally altered two crucial bases of production: land and water. Let us examine the condition of land.[6] In 1990, India had a land area of 328.7 million hectares (ha), 55.6 per cent of which was considered arable. 45.68 million ha or 25 per cent of arable land is irrigated. 22.4 per cent of India's land mass is designated as 'forest'. According to a pioneering study by B.B. Vohra, who was the first to estimate the extent of degradation of India's land, only 42 per cent of the area designated as 'forest' is actually under adequate tree or grass cover; the rest is more or less completely devoid of vegetation (Vohra 1980: 3). With satellite imaging, the government has been compelled to be more accurate in its estimates of deforestation. According to the National Remote Sensing Agency, forests covered 55.5 million ha (16.89 per cent of total land area) in 1972–5. During 1980–2, this area dropped to 46 million ha (14.1 per cent of total area). In the 1990s forests cover a scant 32.8 million ha or 10 per cent of the total land area. Whereas four million hectares of forest area was 'lost' between 1951–76, over the last fifteen years 22.7 million hectares of forest have been cut down (Gadgil and Guha 1992: 196; UNDP 1992: 173).

On the basis of information supplied by the ministry of agriculture, Vohra estimated that more than three-fourths of our agricultural land is degraded due to serious soil erosion, waterlogging and salinization (Vohra 1980: 4). Other sources estimate that 20 million ha or almost 11 per cent of our agricultural land is severely affected by salinization; another 7 million ha have had to be

---

[6] *The State of India's Environment 1984–85: The Second Citizens' Report* contains an excellent account of the status of India's water resources (CSE 1985: 27–48).

abandoned due to salt accumulation (Gore 1992: 111). The productivity of land is diminishing and its use in the future is uncertain. According to one estimate, India loses 6 billion tons of topsoil every year (Gore 1992: 120). While soil erosion is a result of deforestation, excessive grazing, and cultivation of hill slopes without terracing or bunding, waterlogging is usually caused by canal irrigation in poorly-drained soils. It takes nine hundred years to form one inch of topsoil; it may take only one monsoon or one badly-designed canal system to lay it waste forever.

Deforestation, together with the emphasis on building embankments and dams, has led to a steady increase in the incidence of floods in the fertile plains of north India. Annual flood damages increased nearly forty times from an average of Rs 60 crores per year in the 1950s to an incredible Rs 2307 crores a year during the 1980s. The area affected by floods shot up from an average 6.4 million ha a year in the fifties to 9 million ha a year in the eighties (CSE 1991: viii). Ironically, embankments and dams were constructed in order to control the damage caused by floods. Instead, they have prevented the nutrient-rich silt carried by rivers from being deposited in the soil, thereby depriving flood plains of a valuable source of fertilizer. The sediment now accumulates on the river bed, raising it so that the river in spate overflows its sides and devastates more land, lives and property.

None of these losses figure in our national income accounts. When we compute our Gross National Product or our rate of industrial growth, costs such as the loss of topsoil and wasted land are not included in the calculus of economic decision-making. Even though environmental 'depreciation' fundamentally affects the stream of value derived from nature in the future, even though the immediate effects of ecological destruction are real and crippling, these costs tend to remain invisible. In fact, paradoxically, environmental destruction appears on the credit side of the national ledger if it provides a one-time increase in production, even though that increase may destroy all possible future benefits, and may have disastrous 'side effects' in the present. For instance, deforestation will increase GNP through the sale of timber, but there is no enumeration of the losses incurred by cutting trees — the adverse ecological effects or the loss of other use values derived from a forest.

The affluent (usually urban) elite are mostly unmoved by the irreparable loss of national resources. With their power to buy their way out of any crisis by cornering resources for themselves, they have been able to insulate themselves from ecological shock and have even enhanced their lifestyle. This class, which has precipitated land degradation by its extraordinary powers of ownership, control and consumption, even today tends to dismiss the environmental crunch as the gloomy prognosis of pessimists. But the ecological crisis is not some distant doomsday scenario, it is here today in the lives of the poor, experienced as worsening conditions of subsistence. The increasing difficulties that poor rural women and children face in gleaning fuelwood, finding fodder, and fetching water have been extensively documented (CSE 1985: 172–88). Less studied is the steady migration of dispossessed rural people, compelled by ecological degradation to leave the land. According to the UNDP, around 750 million of the world's poorest people live in rural areas. Of these, around 20 to 30 million move each year to towns and cities. And an increasing proportion of these migrants are 'environmental refugees' whose land is so eroded or exhausted that it can no longer support them (UNDP 1992: 58). As ecological destruction disproportionately affects the poor, the *State of India's Environment Report* began by clarifying the politics of the crisis by declaring that 'environmental degradation and social injustice are two sides of the same coin' (CSE 1982).

Ecological Crisis and the Model of Development

Why has the industrialization and urbanization model failed to improve the lives of the poor? Why has it systematically impoverished the natural resource base upon which they depend? These issues can be understood by examining the pattern of state intervention in the allocation of resources. As discussed above, the independent state has been primarily moved by the desire to safeguard and further its own interests and those of its allies: capitalists, merchants and industrialists, and rich farmers:

The influence of the capitalists was reflected in the massive state investments in industrial infrastructure — e.g. power, minerals and metals, and communications, all provided at highly subsidized rates — and in the virtually free access to crucial raw materials such as forests and water.

Large landowners, for their part, ensured that they had an adequate and cheap supply of water, power and fertilizer for commercial agriculture. Finally, the bureaucrat-politician nexus constructed an elaborate web of rules and regulation in order to maintain control over resource extraction and utilization. In this manner, the coalescence of economic interests and the seductive ideology of modernization worked to consolidate dominant social classes. This strategy willingly or unwillingly sacrificed the interests of the bulk of the rural population — landless labour, small and marginal farmers, artisans, nomads and various aboriginal communities — whose dependence on nature was a far more direct one (Gadgil and Guha 1992: 185).

The orientation of the government to look after its own is clearly visible in its policies. Gadgil and Guha (1992), in their excellent analysis of forest policies, show how forests are being mismanaged in order to maximize immediate profits for the state and industry, with complete disregard for the future. The traditional rights of rural communities who use the forests primarily for subsistence have been severely curtailed and reduced to grudgingly granted concessions. The inequalities in access to resources are exacerbated by strategies such as the differential pricing of forest produce: in Karnataka, while bamboo was supplied to paper mills at the cost of Rs 15 a tonne, it was sold to basket weavers and other small bamboo users in the market at Rs 1200 per tonne (CSE 1985: 368).

The tendency of the government to consolidate its power over resources is also shown in its support for large, centralized irrigation and energy projects. These have encouraged the profligate use of natural resources — for instance, in irrigated agriculture where farmers shift to remunerative, water-intensive crops such as sugarcane, uncaring of the ecological or social consequences. Canal irrigation is a highly subsidized system for providing water; while the costs are borne by the state, the benefits mainly go to landowners, further increasing social inequalities.[7] The provision of abundant water for the few through costly irrigation schemes results in induced scarcity for the many. In Maharashtra, people living in the catchment area of a large dam are prohibited by law

[7] The concept of equitable distribution of water rights among the landed as well as the landless is still not widely accepted in India, even though it has been applied successfully in places like Ralegan Sidhi, Maharashtra (Pangare and Pangare 1992) and Sukhomajri, Haryana (Chambers et al. 1989: 155–6).

from using more than 15 per cent of the total available water (Sharma 1990: 230–1). Not only is water directly diverted to the powerful, the opportunity cost of this investment is embodied in the number of small, decentralized schemes to provide protective irrigation to dryland farmers which never materialize because of lack of funds (Sengupta 1993). Considering that 75 per cent of India's arable land depends only on rainwater, the emphasis on irrigated agriculture has concentrated funds on a privileged minority.

Ironically, the bulk of development policies, justified in the 'national interest', actually *diminish* poor people's ability to control and gainfully use natural resources. Every 'national' project is presented as beneficial for the masses even though it requires some poor people to surrender their land or their livelihood. While the 'greater good of the nation' appears to be a laudable cause, it must appear suspicious to the rural poor who are consistently chosen, time and time again, to make all the sacrifices, while those more powerful reap the benefits. Suresh Sharma describes that when Prime Minister Jawaharlal Nehru reassured those displaced by Rihand dam in Singrauli in 1961:

People felt that their suffering would not be in vain. Their instinctive sense of nobility was stirred when Nehru spoke of the Nation and Development. They believed in his promise of a future of plenty to be shared by all. And they half accepted the trauma of displacement believing in the promise of irrigated fields and plentiful harvests. So often have the survivors of Rihand told us that they accepted their sufferings as sacrifice for the sake of the nation. But now, after thirty bitter years of being adrift, their livelihood even more precarious, they ask: "Are we the only ones chosen to make sacrifices for the nation?" (S. Sharma 1992: 78).

In the event, it is obvious that the 'national interest' is merely the interest of the state, industry and rich peasantry.

The model of development established since independence has fundamentally altered the way in which different social groups use and have access to natural resources. The changes wrought by the independent state have created conflicts over competing claims to the environment. These conflicts range from the incessant battle between the forest department and local communities, to the war raging between mechanized trawlers and traditional fishing boats in India's coastal waters, to the controversy over the Dunkel Draft and rights to genetic resources. These claims are not merely for a greater share of the goods, but involve different ways of valuing and using nature — for profit or survival, or some combination of the two. They also involve different worldviews — one driven by the desire to dominate and exploit nature and humanity, the other moved by empathy and respect, sometimes reverence, for the two.

An attempt to understand these relationships between environment and development led to the creation of an Ecological Marxist ideology which marries a concern for conserving natural resources with the issue of social justice. As the *State of India's Environment Report 1982* argues, the future of the natural world is best assured in the hands of the poor and the exploited. Not only will such a transfer of power result in a more just world, it will lead to a husbanding of resources and their utilization for the common good. Efforts to bring about socially just and ecologically sound 'sustainable development' must be seen both as a response to, and a departure from, the shortcomings and contradictions in the present model of development.

II. Development, Environment and Tribal Resistance

NEW DELHI. 28 Jan. 1992. The Chief Minister of Gujarat, Mr Chimanbhai Patel, asserted that work on the Sardar Sarovar Project was proceeding apace. Mr Patel was in Delhi to receive the Independent Review team sent by the World Bank to evaluate the environmental and rehabilitation aspects of the Project. In a meeting with the press, Mr Patel said that any objective appraisal of the Project would show that it was scientifically planned to harness the river Narmada to permanently solve the water problems of Gujarat. The government of Gujarat had designed a truly enlightened rehabilitation package for dam-oustees. Opposition to the Project was limited to some misguided environmentalists. (Press Release: Government of Gujarat.)

I was in Anjanvara, a village of Bhilala tribals in the area to be submerged by the Sardar Sarovar dam. It was late in the evening and I was sitting outside the hut of the *pujara* (priest), watching him roll a bidi. We were talking about gods. The pujara said, 'People think of Narmada as bigger than all the gods, bigger than the earth. She can grant all that anyone desires.' 'Then why aren't you better off? Why doesn't she give *you* all that you want?' I asked. 'But we live in the belly of the river,' said the pujara. 'Sometimes she listens to us and sometimes she doesn't.'

The Model of 'Development'

The attempt to achieve modern industrial growth has been based on two interrelated processes: one, the unchecked use of the earth's natural resources; and two, the transformation of people, often against their will, into a dispossessed working class. These processes were not new; they had their antecedents in India's history of colonial and precolonial extraction, and they continued after Independence, though they were legitimized in different ways. During the colonial period, modernization was part of the imperial mission of civilization and improvement of the natives — the white man's burden (R. Guha 1988). For the independent state, modernization was essential to the project of national development. As we saw in the first part of this chapter, the ideology of 'national development' has been used to legitimize exploitation.

The project of national 'development'[8] is not limited to the Indian state alone, but is embedded in contemporary global structures such as the arrangement of the world into nation states, and the expanding system of international capitalism. While concentrating on the relations between the Indian state and its subjects, we must remember that these relations and those of international political economy mutually shape each other. The model of development as modern industrial growth was derived from the historical trajectory of former colonial powers such as Britain, France and Germany — a model that newly independent states sought to emulate. The pursuit of growth necessitated large injections of capital into the national economy for developing industrial infrastructure — an investment that has often been financed by foreign funds. A typical instance of such state expenditure aimed at fostering economic growth is the Narmada Valley Project in western India. This is a gigantic scheme to harness the waters of the river Narmada for irrigation, power generation, and drinking. The Project has been partially funded by the World Bank, and partially by bilateral aid.

[8] I have placed development within quotation marks to signify that the positive connotation of the term is an ideological construct and should not be accepted uncritically. However, this is a clumsy device, so I shall not use it through the rest of the text, merely adjuring the reader to keep in mind that the quotes, though invisible, are present.

Ironically, such external borrowing for capital-intensive projects has served to increase indebtedness to the point that it has undermined the very objective of national development. The easy availability of credit for development in the 1960s and 1970s has, in the last decade, become a millstone around the neck of developing nations. Foreign creditors rescheduled debt repayment for borrower nations on the condition that they would undertake 'structural adjustment' — trade liberalization and reduced spending on public welfare. The austerity measures thus entailed have left the poor tightening their belts over their hungry bellies (George 1988). Being in the red has further shifted the orientation of national governments from meeting domestic needs to a preoccupation with debt servicing. In order to give creditors a guarantee of its ability to maintain the schedule of repayment (a condition for new loans), the state has to play an increasingly repressive role, keeping the working classes in line and preventing social unrest (Canak 1989). Thus, the state's indebtedness to foreign capital, incurred in order to develop, has today become a constraint, reducing the options available for autonomous growth.

In the name of development, national elites, through the institutions of the state and the market, and often in collaboration with foreign capital, have appropriated natural resources — land, minerals, forests and water — for conversion into commodities. The circulation of goods which this has brought forth has taken place primarily among the already affluent owners of capital and other elites. Elites, who have the desire and the power to profit and consume, have thereby impoverished the earth of its natural wealth and, through degradation and pollution, have rendered it unlivable for future generations. This has been called the second contradiction of capitalism — a contradiction between the ever-expanding circulation of capital which has no limits and a Nature which has many (O'Connor 1988).

The earth's impoverishment has meant that communities who depend on the natural base for sustenance have been deprived of their resources. This alienation cannot be adequately described in terms of the loss of a material livelihood alone; it is most profoundly a wider loss of cultural autonomy, knowledge and power. In the name of development, people have been pushed off the land; their forests and water have been taken over by the state and the market, so that they have been deprived of everything except their labour power. Just as nature has increasingly become a commodity, so has human labour. This is the first contradiction of capitalism — that labour works with nature to produce value, only to have it appropriated by the owners of capital. This unequal relationship has been enforced by the authority of the state, based ultimately on its monopoly over violence (Tilly 1985). The coercive aspect of state power has been concealed behind the institutions of liberal democracy. Thus, in another ironic twist, human impoverishment has occurred even as people have participated in the political processes of democratic decision-making, apparently consenting to their own ecological, economic and political marginalization.

Ecological Marxism

Such a theoretical understanding of development, as a process that exploits the labouring classes as well as nature, has come about fairly recently. Traditionally, Marxist analyses of social conflict have paid much more attention to the conflict between labour and capital, or the social relations of production, than to the exploitation of nature. The incorporation of the capital/nature contradiction into Marxism is the contribution of Ecological Marxists such as James O'Connor, Watts, and others, who have tried to bring ecological concerns into a Marxist analysis of political economy. This interpolation is not uncontroversial; some scholars have argued that 'red' and 'green' agenda cannot be mixed in this way. Marx himself was uncritical in his appreciation of capital-intensive technology and could not see the environmental havoc that industrialization entails. Therefore the attempts to find an authentically ecological outlook in Marx's philosophy of nature cannot be successful (Clark 1989).

The neglect of an ecological perspective in Marx's own work is not an oversight. Marx was conscious of the Malthusian argument that natural limits such as a slowly growing food supply would act as a brake on social betterment. But he concentrated on rebutting Malthusian ideology by adopting an essentially Ricardian view of *social* rather than *natural* limits to capitalism. That is, Marx chose to stress that the *social* relations of production need to be transformed to fully realize the fruits of technological development. Therefore, Marx's core concept of the labour process underrepresents the significance of natural conditions which cannot be manipulated and overrepresents the role of human intentions and powers for transforming nature (Benton 1989: 64).

But at a deeper level, Marx implicitly supports environmentalism by opposing commodification — a process which divorces all products (fashioned by human labour working with nature) from their intrinsic worth, and reduces them to a common economic matrix. Marx's critique of capital depended fundamentally on his critique of commodity fetishism, which is an epistemological critique of economism and development, with their corollary of environmental destruction. In this more profound sense, Marx anticipated the class and nature problematic.

Resistance to Development: Party Politics *versus* Social Movements

The process of development has not been easy or straightforward; it has been continually contested by competing groups of elites, within and between nation states (Evans and Stephens 1988). More critically, the attempts of elites to exploit in the name of development have been challenged and collectively resisted by the very people that they have sought to marginalize. Popular protest has occurred at many different levels, and has had many different objectives — from retaining access to natural resources, to getting a fairer deal in the work process, to asserting cultural autonomy — but it has been ultimately unified by its attempt to decentralize power into the hands of the exploited majority so that they have more control over their lives.

In India, among the many struggles against national 'development', one has received increasing scholarly attention in the last fifteen years — resistance in the form of social movements. Even though social movements tend to be small and localized while trade unions and peasant parties have large followings, it is the former which have become the subject of scholarly treatment. This academic shift, similar to that which occurred in Europe, is due as much to the resurgence of this form of protest as to the intellectuals' own ideological disillusionment with electoral politics and mainstream socialist politics. Kothari argues that these movements are 'really to be seen as part of an attempt at redefining politics at a time of massive attempts to narrow its range, different from electoral and legislative politics which has relegated large sections of the people outside the process of power' (Kothari 1988: 46).

In part, intellectuals seem to have surrendered their interest in state power as an arena of struggle because of the failure of progressive mass-based parties to form stable governments at the national level. Several reasons are offered for this: party politics is corrupt and compromised; it contains inherent tendencies towards centralization; in any case, new institutional spaces are opening up at the grassroots. By espousing the cause of new social movements, intellectuals have filled an analytical void created by their perception of the failure of bourgeois democracy.

New Social Movements in India and Europe

The surge of Indian scholarly interest in social movements parallels that which has occurred in western Europe at the same time, around Green movements — a number of closely related actions on the issues of nuclear disarmament, peace, feminism, and the environment[9]. Intellectuals discussing these movements make claims similar to those made by Indian scholars about their politics: 'the sphere of action of the Green movements is largely a space of non-institutional politics which is not provided for in the doctrines and practices of liberal democracy and the welfare state' (Offe 1985: 826). For their difference from mainstream politics, these movements have even been called 'antipolitics' (Berger 1979).

In the European context, Green movements are perceived to be a departure from the orthodox Marxist concerns about the centrality of work and class relations in defining the lines of social conflict. As Rudolf Bahro said, 'some comrades, whose way of thinking is fixated on a past era that cannot be brought back, speak staunchly of a still too low level of class struggle. You fail to see that the emperor has no clothes. The world-historical

---

[9] While all these issues have separate and noble lineages of struggle which long pre-date the Green movement, the new social movements claim to be new precisely because of the way in which they have woven these different strands of social action into one unified agenda.

mission of the proletariat was an illusion . . . ' (Bahro 1982: 57).
He goes on to say that the emergence of the ecology movement
signals that countless people who are not mobilized by specific
economic class interests proclaim and organize their resistance in
diverse ways (Bahro 1982: 107).

Intellectuals of Indian social movements share with their
counterparts in the European Green movements a critical unease
about party politics, a wariness that has been expressed in self-
conscious deliberations about the relationship between parties and
movements. Apart from this similarity, can these social movements be seen as 'new' or 'antipolitics' in the Indian case? Unlike
Bahro's emphasis on the absence of concerns about class issues in
the Green movements, social movements in India continue to
centre on the conflict over production relations. While the
European experience has been called a shift from 'red' to 'green'
— that is, from Marxism to environmentalism, the Indian political process seems to be tinted in both hues. The struggle over
nature, for example, has an inherent class dimension because
nature also provides resources which are the bases of production.[10]
Unlike Europe, where ecological crisis is perceived equally as a
threat to biological survival in global terms, in India conflicts over
nature tend to closely follow the battle lines between those who
produce and those who own the means of production (Guha
1988: 2578).

However, in one respect, scholarly interest *has* moved away
from the consideration of one set of productive resources to
another which had earlier been peripheral to the concerns of
political organizations. From an examination of conflict in the
factory and the field, intellectuals have moved to study conflict
around forests and rivers — a shift which is sometimes perceived
to be a move towards a 'green' agenda of ecological sustainability.
As may be apparent, the Indian exercise of defining what is
'environmental' and what is not, is somewhat arbitrary. Whereas
conflict around agricultural land tenure and use, or around the
workplace, is usually not treated as environmental, conflict
around forests or water resources is deemed to be quintessentially

[10] Of course, as O'Connor points out, the meanings attributed to nature
are not exhausted by its use as natural resource; ecological politics are about
'class issues even though they are *more* than class issues' (O'Connor 1988: 37;
emphasis in original).

environmental. This distinction is hard to maintain because land management is also of critical ecological importance, and confrontations over forests are also about the ownership and control of the means of production. The unexamined presumption that conflict over forests and water is environmental and that over agricultural land is not, stems, I would speculate, from the class background of the scholars who make these classifications, and who tend to see forests and rivers as 'wilderness' (and therefore 'nature' and 'environment'), and not primarily as sources of livelihood. That is, an upper-class view of what is environmental prevails. For most rural communities, there seems to be a continuity in the way in which land, forest, water and other resources are regarded, primarily as sources of subsistence.

In the Indian context, the inseparability of, and continuity between, the 'red' and 'green' agenda is asserted by Ramachandra Guha, who observes that the material structure of society serves as the 'landscape of resistance' for people fighting against exploitation. While social relations and forces of production limit the forms a culture (and within it, resistance) may take, our analysis

has to include not only the economic landscape but also the natural setting in which the economy is embedded. For, while production relations sharply define the boundaries of political structures and cultural systems, they are in turn limited by the ecological characteristics — the biota, topography, and climate — of the society in which they are placed (Guha 1989a: 5–6). While Guha's notion — that ecological specificities limit and modify social relations — is an important corrective to theoretical approaches that ignore the environment, by itself this does not constitute a 'green' agenda. Environmentalism is not concerned simply with *nature* per se, it is concerned with the *sustainable use* of nature. It is generally thought that sustainability has been the attribute of economies organized for subsistence, as opposed to profit. The low-intensity use of nature to meet basic needs is ecologically sustainable, for it maintains the regenerative capacities of natural resources. Thus, the struggles of rural communities for retaining control over natural resources for subsistence can be interpreted as embodying a 'green' agenda of environmentalism.

The Ideological Dimensions of Social Movements

Ecological conflicts in India are continuing struggles over production and extraction. The history of forest conflicts, for instance, has been a long battle between the state which sought to incorporate the forest into a profit-economy, and local communities who fought to retain its place in their subsistence economy (Gadgil and Guha 1989). However, collective action over natural resources not only raises questions about ownership and control, it is also a claim about different *relationships* with nature. As E.P. Thompson said, 'Every contradiction is a conflict of values as well as a conflict of interest.' New social movements too are seen as formulating a far-reaching critique of the ideology of development, rejecting its claims that the ever-increasing exploitation of humans and nature constitutes progress, and reshaping people's ideas about what is socially good and desirable. Besides the contest between two versions of economics — the political economy of profit and the moral economy of need — this critique is believed to extend to a repudiation of the underlying cultural values that are privileged by the ideology of development — the ecological hubris of trying to attain mastery over nature, the economism that ranks profit over all else, and the pursuit of technological expertise which dominates over other, more egalitarian and 'organic' ways of knowing the world. This harks back to European claims that Green movements are not sparked by '*problems of distribution*, but concern the *grammar of forms of life*' (Habermas 1981: 33; emphasis in original).

According to Ramachandra Guha, in India the environmental critique of the values of development builds upon three ideological streams (Guha 1988). The first, which he calls 'Crusading Gandhian', upholds the pre-capitalist and pre-colonial village community as an ideal of social and ecological harmony. It criticizes the domination of modernist philosophies such as rationalism and economism, and propagates an alternative philosophy which has roots in Indian tradition. The second stream, called 'Appropriate Technology', is influenced by socialist principles but reconciled to industrial society. It emphasizes the liberating potential of resource-conserving, labour-intensive technologies. The third stream, that of 'Ecological Marxism', holds that political and economic change must be prior to ecological concerns, and collective action aimed at systemic transformation must come first. This stream is philosophically at odds with Gandhian values because of its overall faith in the emancipatory potential of modern science and technology. While conceptually separate, these ideologies have been influential in modifying each other. The Gandhian critique of modern science has muted the celebration of technicism by the Marxists, while the Marxist analysis of exploitation has compelled the Gandhians to incorporate a more egalitarian perspective on social change. The result of this cross-fertilization is an environmentalism which builds upon the strengths of these different ideologies to construct a far-reaching critique of the values underlying ecologically destructive development.

Another stream of Indian environmentalism, although not a critique of development per se, is that of preservation. One of the earliest and most celebrated instances of preservation was Silent Valley in south India where the central government overrode the protests of the state government and local people and cancelled a proposed hydroelectric project in a pristine tropical rainforest (Herring 1991). Prompted by the international call to save biodiversity, and the urging of preservationist groups such as the

Bombay Natural History Society, the Indian government has made several 'wilderness' areas off-bounds to local users by designating them as national parks. While the creation of parks and wildlife sanctuaries *has* facilitated environmental conservation, this is preservation by fiat, for local people's rights have usually been overlooked and they have rarely been involved in management. This has also contained environmental action to a sphere of low overall impact for, through its other policies, the government has continued to pursue the environmentally destructive goal of industrial growth. Preservation has tended to be a token effort, creating islands of endangered wildlife which are perpetually encroached upon, resented by the people who live near them[11].

Indigenous People as Embodiments of Resistance

Guha's categorization of the ideological bases underlying Indian environmentalism can be extended to include another stream, similar in its values to the Gandhian, but rooted in a different tradition: the belief that the cultural beliefs and practices of 'indigenous communities' constitute a critique of ecologically destructive development and provide an alternative vision of a sustainable human-nature relationship.[12] It is claimed that over centuries of living sustainably with nature, adivasis have acquired a deep knowledge and understanding of ecological processes so that they are ideal natural resource managers (Shiva and Bandyopadhyay 1990: 77). Thus, scholars of social movements say that 'rural women and indigenous people . . . still retain the *aranya sanskriti* [forest culture] which is based on the creative interdependence

[11] A classic case of a conflict between park officials and villagers occurred in November 1982 when people living around the Bharatpur bird sanctuary were barred from exercising their traditional rights to graze cattle inside the sanctuary. The subsequent clash resulted in police firing that killed one person (Prasad and Dhawan 1982).

[12] Like 'environment', 'indigenous' is hard to define. The term is problematic outside its original context of the Americas where, historically, there has been a sharper differentiation between 'natives' and European settlers. The use of 'tribe' as an alternative is also difficult because of the porosity of the boundary between 'caste' and 'tribe', both of which have existed side by side for centuries in India (see Chapter 4). I shall avoid controversy by using *'adivasi'* (literally 'original dwellers'), a widely accepted Indian term; Hardiman presents a persuasive case for this usage (Hardiman 1987b: 11–16).

between human evolution and the protection of forests (Parajuli 1991: 179). Vandana Shiva and J. Bandyopadhyay reiterate this theme when they speak of the cultural lessons of diversity and democratic pluralism learnt by Asian societies 'modelled on the forest'. They proceed to, claim that

the forest as the source [of life] also means that forests and trees must be treated as sacred. The sacred is inviolable: its integrity cannot be violated. If Asian civilizations have survived over centuries it is because they learnt to be like the forest, sustaining both the forest and the culture through time . . . For these cultures, all life, both human and non-human, is in symbiosis. Human society is not predatory but in rhythm with the forest (Shiva and Bandyopadhyay 1990: 67, 77).

Overlooking the glib biological analogies here, informed as they are by functionalist notions of homeostatic social systems, this basic view that adivasis are sustainable managers has gained wide acceptance today among intellectuals writing about development and resistance. The practices of adivasis are said to exemplify 'the life-enhancing paradigm' (as opposed to the modern 'life-destroying paradigm') where *renewability is the primary management objective* (Shiva and Bandyopadhyay 1990: 74; emphasis in original). It is declared that

. . . all serious studies of natural resource use by indigenous peoples show that their traditional ways of life have been brilliantly conservationist . . . [T]heirs is an ecological wisdom that is intricately woven into the very fabric of their cultures; for the most part it is not an articulated, conscious 'body of knowledge' . . . [T]heir entire way of life expresses an ecological wisdom that enables them to take care of their forest environment (Taylor 1990: 184).

Thus, indigenous ways of knowing, which people are unable to articulate or even be conscious of, are expressed 'on their behalf' by intellectuals.

This view of the adivasis as ecologically noble savages is not unique to India; it is frequently voiced across the world among conservationists concerned about saving the forest. In Europe, the noble savage was first the idealized vision of Rousseau, Thomas Moore, and others, of the inhabitants of the New World. The belief that native Americans 'lived in close harmony with their local environment' was resurrected in this century among conservationists. Paralleling Shiva and Bandyopadhyay's conceptualization, the modern world was seen as being divided into 'two systems, two different irreconcilable ways of life: the [Native American] world — collective, communal, human, respectful of nature, and wise — and the western world — greedy, destructive, individualist, and enemy of nature' (from a report to the International NGO Conference on Indigenous People and the Land 1981, quoted in Redford 1991: 46).

The belief that adivasis in forests everywhere are conservationist also recurs among north American Deep Ecologists, who see their philosophical principle of 'biocentrism' (as opposed to 'anthropocentrism') realized in eastern religious traditions and, at a more popular level, by 'primal' indigenous people in non-western settings who, through their material and spiritual practices, subordinate themselves to the integrity of the biotic universe they inhabit. The coupling of (ancient) eastern and (modern) ecological wisdom seemingly helps consolidate the claim that deep ecology is a philosophy of universal significance (Guha 1989b: 73–6). In India, the view that 'traditional' religious worldviews depict the innate ecological sensibilities of Indians is found in the writings of Vatsyayan (1992) and Banwari (1992), who interpret Vedic rituals and myths as embodiments of environmentalism.

According to this environmentalist view, in the knowledge and belief systems of the adivasis lie our hope for the future. The wisdom of 'indigenous' people is thought to have contemporary relevance, for it is believed to be inherently ecologically sound, as proved by their sustainable survival strategies (Redclift 1987: 153). This wisdom, then, forms the philosophical foundations of present-day social movements of people who resist and challenge the dominant ideology of development. 'In the third world today the alternatives [to development] are often there in the present, surrounding the islands of barricaded modernity' (Nandy 1987: 88). It is claimed that adivasis, who have been marginalized by development, can mount a thorough going critique of it and, through the example of their ecologically wise culture, present an alternative vision of sustainable, ecologically respectful living with nature. Such a cultural critique and alternative is not merely an abstraction; it is realized in the present through social movements. Thus, 'new social movements are . . . sites of creating and regenerating [the] subjugated knowledge . . . [of] indigenous people, women, and other marginalized groups . . . ' (Parajuli 1991: 183).

The collective resistance of indigenous people is not a rearguard action — 'the dying wail of a class about to drop down the trapdoor of history' — but a potent challenge which strikes at the very heart of the process of development.

The Formulation of an Agenda for Research

I formulated my research hypothesis after being influenced by the preceding theories. I expected that adivasis acted politically in response to their experience of development — a process which has resulted in the alienation of their natural resource base and their subsequent cultural impoverishment. That is, they come together in a social movement to collectively resist the appropriation of their resources by state and market. The conflict between social movements and the state is not simply that of differing interests but, equally profoundly, that of differing values. As people not fully incorporated into the state and the market, adivasis can draw upon traditional values such as reverence for nature as the source of their cultural critique of development and the basis of their resistance. This resistance is not merely reactive; through it, adivasis construct a creative alternative to the dominant and destructive system of development, based on their tradition of living sustainably with nature.

The ongoing struggle of the adivasis in the Narmada valley in central India seemed to be a living example of the resistance of 'indigenous' cultural communities to development. The Indian government plans to dam the river Narmada, harnessing its water for irrigation and power generation. The reservoir of the dam will submerge an area of forested hills, displacing the adivasis who subsist upon this environment. Their fight against displacement appeared to be intrinsically an environmental movement, for did they not worship nature and use it sustainably? While the dam was both a part and a symbol of development, the movement against the dam seemed to embody cultural resistance and an alternative to development. This alternative extended to the very mode of political action in which adivasis engaged — decentralized, grassroots mobilization which challenged the authority of the state to act 'on behalf' of the people. The Narmada Bachao Andolan (Save Narmada Movement) represented the marginalized, uncorrupted 'alternative political culture' of the adivasis. I intended to go to the Narmada valley and, by living with the adivasis, discover their relationship with nature, how it changed with their experience of development (which included the dam), and their struggle to create an ecologically sustainable and socially just alternative world.

On going to the Narmada valley, I discovered a somewhat different reality. I had earlier believed that the movement against the dam was the only way in which adivasis acted politically; I found that they were also organized into a Sangath (union) which fought for adivasi rights to land and the forest. Moreover, this did not exhaust adivasi politics; people participated with enthusiasm and great energy in waging and settling village-level feuds about honour. Fieldwork revealed that there were several different levels of politics, and that it was essential to incorporate all of them into the study in order to better appreciate adivasi life.

More disconcerting, though, was the rapid discovery that adivasi life was not at all what I had imagined it to be. My expectation that I would encounter a community which lived in harmony with nature, worshipping it and using its resources sustainably, turned out to be both true and false. Therefore my neat theoretical framework linking nature–culture relationships to political critique, action and change, crumbled into an untidy jumble of contradictions. The dissonance between my romantic notions of adivasis, based on scholarly writing on the subject, and the everyday lives of adivasis, led me to ask the questions that run through the subsequent pages: What is the adivasi relationship with nature today? What is their relationship with the state? How do people, whose struggles are the subject of theories of liberation and social change, perceive their own situation? How accurately does the environmentalist critique of development represent the lives of the people who are thought to be at the forefront of environmental movements? While these issues have been analyzed in relation to the Bhilalas of Alirajpur who live in the submergence area of Sardar Sarovar dam, they bear upon the lives and struggles of adivasis elsewhere, who share a past and a present of development and resistance.

3

A History of Adivasi–State Relations

Now I turn to the emergence of the hill adivasis at the conjunction of local and global history, tracing the changing relationship between adivasis and the state as they struggle for control over natural resources and examining how this conflict transformed both state structures and adivasi communities. While nature has consistently been the ultimate prize for which battles, both overt and invisible, were fought, I shall trace the *variations* in the appropriation of nature. As the extractive and executive powers of the state increased, so did the sway of markets, both transforming nature and the processes and agencies by which nature was put to social ends. The expansion of the state and the market also transformed people's understanding of their predicament and affected their actions. By analysing adivasi resistance to external domination, I situate the constitution

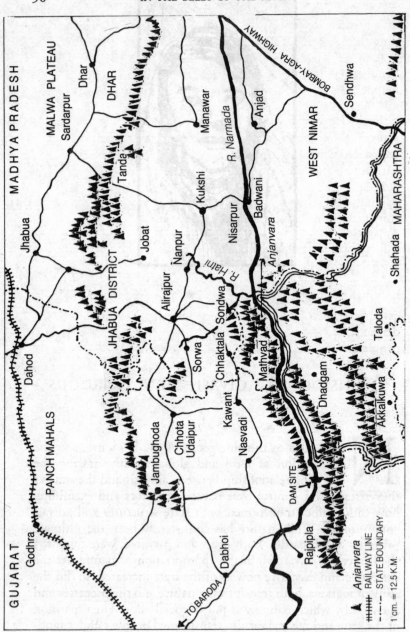

Fig. 2: Present day map of the area around Alirajpur indicating Anjanvara

of adivasi community and identity in the context of, and in opposition to, the expansion of state and market structures. By looking at consciousness as the problematic fusion of objective conditions and subjective experiences, I sketch some of the inherent contradictions of resistance — a theme that recurs throughout the subsequent analysis.

First a caveat: This reconstruction is based mainly on historical sources pertaining to Alirajpur and Akrani (Khandesh). I did not have access to records for Mathvad, the tiny *thakurat* (minor estate) within which Anjanvara lay (see Figure 3). Therefore this is a general history of the region, informed by accounts of changes in adjoining areas. It must be borne in mind that historians consider Alirajpur a 'remote' area (Hardiman 1987a: 32) and, within that, Anjanvara and other villages in the submergence area seem remoter still. So the impact of outside groups has been more diffuse than in the rest of India. As we shall see, 'remoteness', i.e. the distance created between Anjanvara and administrative-commercial centres by hills and forests, has been crucial in limiting impact. Yet, in extrapolating Anjanvara from the records of the literati, our understanding of change is strongly skewed in favour of the views of people in power within urban-administrative centres. This account, like others based on documentary sources, will probably err on the side of overemphasizing externally-induced change.

Another difficulty in writing such a history is that most early accounts of Bhil and Bhilala regions tend to be ahistorical, cataloguing the 'manners and customs' of the natives. This tendency still prevails among modern Indian writers. For example, S.C. Varma's book *The Bhil Kills* is primarily a compilation of ethnographic odds and ends (Varma 1978: 22ff). The Dhar gazetteer of 1984 reproduces verbatim the ethnographic writing of Russell and Hiralal from 1916, without finding it necessary to update that record. The tribal, as 'the primitive living in the forest since time immemorial', is denied history. Where there *is* some record of change, it is merely the transfer of kingship; history is but a chronicle of conquest.[1] So delineating a social history requires as much imagination as information.

---

[1] There has been very little research into tribal political action in this region. Ghanshyam Shah and Arjun Patel (1993) have reviewed the literature on tribal movements in Gujarat, Maharashtra and Rajasthan; such a compilation is long overdue for Madhya Pradesh.

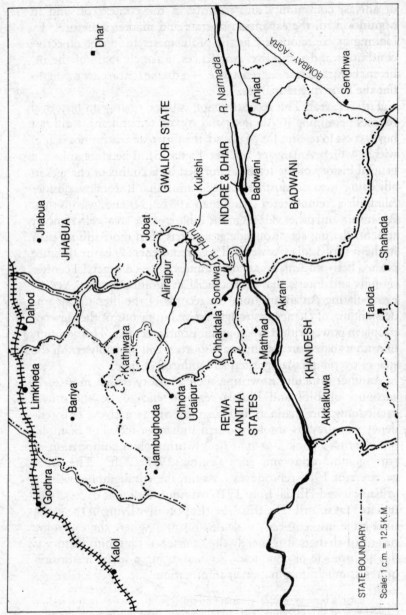

Fig.3: Mathvad and Adjoining States, *circa* AD 1800

The region of which we speak lies in the south-west corner of Madhya Pradesh in western India. Following the course of the Narmada after Hoshangabad, we pass through the plains of Nimar in Khandwa, Khargone and Dhar districts of Madhya Pradesh. From the flatness of Nimar, the Narmada flows into a rocky valley flanked by two hill ranges — the Vindhyas on its north bank and the Satpuras on the south (see Figure 2). Here the river becomes the border between two states, Madhya Pradesh on the north bank and Maharashtra on the south. This is where Anjan-vara lies, at approximately 22°N latitude and 74° 12'E longitude, in Alirajpur tehsil of Jhabua district, Madhya Pradesh. Just seven villages downstream of Anjanvara, the Narmada passes from Madhya Pradesh into Gujarat, briefly becoming the boundary between Gujarat and Maharashtra. This region, Mathvad, of which Anjanvara is part, is thus separated by the Narmada from Maharashtra to its south, and is bounded on the west by Gujarat's Baroda district, and by Nimar on the east. At the confluence of several different cultural traditions, Mathvad has a linguistic rich-ness and sartorial splendour which can only be called cos-mopolitan.

Being the western-most tip of the Vindhyas, Mathvad is moun-tainous; the hills are covered by dry deciduous mixed teak forests which have been partly cleared for cultivation. The rocky terrain, broken by gorges and streams, has thin soils. The slopes are watered only by the monsoons and rain tends to vary wildly from year to year. While the rainfall averages around 60 cm, in the last seven years it has fluctuated between 41 cm and 96 cm. Conse-quently, drought is a frequent visitor to this land. Mathvad is a part of Alirajpur tehsil which has an area of 2237 square kilo-metres. The region is sparsely settled compared to the national average, with a density of only 87 people per square km. Mathvad is poorly connected by road; of its two roads, only one is partly paved. Consequently, people are used to walking long distances. Alirajpur presents a dismal picture in terms of various human development indicators: a scant 4.6 per cent of the population is literate; only 2 per cent of the women can read and write. Of the tehsil's total population of 196,000, only 14 per cent has access to government medical services. A mere 55 (16 per cent) of the 339 villages in the tehsil are electrified. Most villages have no source of safe drinking water. The population of Alirajpur is

overwhelmingly tribal; almost 89 per cent of the people belong
to the Bhil and Bhilala tribes. Another 6.7 per cent belong to
various scheduled castes. The small non-adivasi population includes Bohra Muslims and Banias, who tend to be traders and
moneylenders, and Rajputs. The town of Alirajpur also contains
two distinct communities — the Ashadha, a former low caste
which claims Rajput status, and the Makranis, former adventurers
who came from Afghanistan and fought as mercenaries along with
the Bhils, who have now settled into more respectable professions.

Pre-eighteenth Century: The Rajput Invasion

Very little is known about this region before the eighteenth century. We can only assume that the area was populated by small
tribal communities who lived in the forest and cultivated small
patches of land which they cleared and burnt. Till the seventeenth
century, the forest was dense and 'the hilly tracts . . . were a great
breeding place for wild elephants . . . [who] were frightened off
by the frequent passage of armed bodies during the Mughal
conquest of the Deccan' (KG 1880: 30).

The elephants were not the only ones to be scared off by the
Mughals. As a result of the Muslim invasion of Rajasthan, Gujarat
and Malwa that occurred around this time, many Rajput warriors
fled these areas and came to settle in the Narmada valley. Bhils,
who had ruled this entire region till the eleventh century, came
to be gradually displaced by the Solanki (Rajput) rulers of Gujarat
(Nath 1960: 13), so that they could retain their rule only in the
hills of the Vindhyas and Satpuras during the medieval period
(Hardiman 1987a: 28). The Rajput takeover reached its greatest
heights in the fifteenth century. Around 1437, the Rathor (Rajput)
chieftain Anand Dev claimed for himself the kingdom of Alirajpur, his kin carving up Phulmal, Sondwa and Jobat as their
territory (Luard 1908a: 598). About 1440, 'the chief of Alirajpur
allowed one Ade Karandeo Parmara, who had ousted the original
Bhil chief [Motia Bhil][2] to hold the estate' of Mathvad (Luard
1912: 68). Many of these Rajput rulers married Bhil women and
the result of that miscegenation is said to be the Bhilalas.[3]

[2] Motia Bhil is still worshipped by the Bhils of Mathvad on Dussehra day.
[3] Historical accounts do not distinguish between Bhils and Bhilalas and, in
this chapter, I am forced to conform to that practice, even though that difference

The Bhil polity that existed before the Rajput influx was not centralized. Territory was divided into a large number of localities, each under a hereditary chief called *naik*. However, there was little to set the chief apart from other Bhils. He could call on the Bhils of his area to serve as bowmen, for defence or to raid the peasants of the adjoining plains, especially in years when drought induced scarcity. But there was no taxation, no system of surplus extraction to make the chief richer than his band (Hardiman 1987a: 29).

According to G.S. Aurora, the Rajputs were unquestionably superior to the Bhils militarily; Rajput hegemony over the Bhils was initially established by the sword (Aurora 1972: 33). Later, as the Bhils were forced to accede to their subjugation, the Rajputs selected some tribal chiefs to act as representatives of their authority in the villages, such that the chiefs came to be the link between their ruling political system and the traditional tribal polity.

However, Rajput hegemony nested within superordinate Mughal domination. During the medieval period, the Mughals not only emerged as the most powerful factor in the Indian power system but also achieved a position of authority at the centre, and became a source of legitimization of authority at the lower levels (Aurora 1972: 33). The Rajput kingdom of Alirajpur chose not to challenge the power of the Mughals and paid tribute to the Badshah at Delhi. Alirajpur adopted Mughal forms of administration in the latter half of the eighteenth century. The kingdom or province (*suba*) was divided into districts (*sarkars*) and sub-districts (*parganas*), creating a hierarchical structure for the collection of revenue. An administrator (*kamasdar*) was appointed to oversee each pargana.

Before the Rajputs, Bhils settled disputes through *panchayats*, where a group of elders from the village community would be both judge and jury. Conflict was negotiated through the *bhangjadya*, the go-between who would work to resolve quarrels. The Rajput rulers bypassed this system of arbitration and chose only to recognize the authority of the village headman or *patel*. The naiks (tribal chiefs) were displaced by Rajputs who were allotted *jagirs* (estates). Despite the adoption of new administrative structures, the tribals were able to maintain a high degree of

is a matter of great importance to the people themselves (see Chapter 4).

political autonomy at the local level by continuing to take disputes to the panchayats and by incorporating many of the local 'outside' leaders informally into the system by granting them 'elder' status.

By the end of the eighteenth century, the segmentary tribal societies lived under the overarching authority of Rajputs who were relatively more centralized in their kinship-based political organization. Political centralization went hand in hand with the establishment of the revenue hierarchy. Curiously, this process of patrimonialism occurred much later in Alirajpur than in, say, nearby Khandesh, where Todar Mal's revenue settlement had been implemented by Shah Jahan in 1634. Why did Alirajpur remain untouched by the claims of kings and armies? Perhaps the area was perceived as being too poor to exploit, so that it, like neighbouring Badwani 'with its barren soil and hilly surface, escaped the covetous eyes of the Muslim emperors . . . ' (WNG 1970: 64).

The Maratha Campaign Against the Bhils

The Alirajpur kings owed allegiance to the Delhi durbar, but by the middle of the eighteenth century, around Alirajpur — in Malwa to the east, Rewakantha to the west and Khandesh to the south — Mughal supremacy was eclipsed by the growing power of the Marathas. While carving out an empire, the Marathas ruthlessly suppressed all resistance, including that of adivasis. Ram Chandra Bhuskute, subedar of the Peshwa (the Maratha ruler of Poona), subdued the Bhils with drastic measures. The rebels were brought to Khargone and ordered to give security for good behaviour. Those who complied were presented with a special collar to wear and those who did not were beheaded on a public *chabutra* (platform) in Khargone. 'The pillar to which the victims were tied and the axe with which they were executed . . . were worshipped as recently as 1931 on Dussehra day as emblems of law and order' (WNG 1970: 59).

The Marathas did not remain united for long. The Peshwas, Holkars and Scindias fought amongst each other for supremacy in Malwa and Khandesh, devastating the land with their constant battles. There was little hope of stable administration in the midst of internecine warfare. And on top of ruin wrought by warring armies and tribal raids came drought. In 1803–4, the rains failed and famine stalked the land. The Bhils took to the hills 'and, from their stronghold, issued forth periodically to loot and lay waste the plains. . . . Nor did travellers pass with impunity through the hills, except in large convoys. By AD 1800, during the disturbed period of the wars between Holkar and Scindia, they became so bold all over the country as to terrorize even the several local chieftains and extract large indemnities from them' (Nath 1960: 13).

These guerrilla raids of the Bhils upon rich plains villages were a thorn in the side of the Marathas. The Bhils were meted out swift and bloody reprisals. In 1804 'at Kopargaon, Balaji Lakshman, tempting from the hills a large body of the Chandor Bhils, surrounded and massacred them.[4] This treachery only made the Bhils fiercer, and the Maratha officers retaliated by most cruel massacres at Chalisgaon, Dharangaon and Antur' (KG 1880: 254). The bloodthirstiness of the Marathas against the Bhils indicates that the exploitation of adivasis long predates India's colonial history.

The Rise of British Power

By the 1750s the British had been in India for more than two hundred years. The subcontinent was full of squabbling princes who hired foreign mercenaries to lead their armies and, on occasion, besought the British for military assistance.

The British took advantage of Maratha disarray to gain a foothold in Malwa and Khandesh (Manohar n.d.). A series of wars between the British and the Marathas, with British skulduggery playing off one Maratha faction against another, culminated in November 1817, when the Peshwa, the Nagpur chief and Holkar rallied to form a united front against the British. This, 'the last great alliance of the Marathas', failed when Holkar's army was defeated in December 1817. Under the terms of the treaty of Mandsaur, made after this defeat, Holkar ceded to the British all his territory south of the Satpuras, including the entire province of Khandesh, which came to be a part of Bombay Presidency (KG 1880: 254).

---

[4] According to Dinanath Manohar, 15,000 adivasis were killed in fifteen months, including the killings at Kopargaon. Manohar's account records the Marathas as being led by the Peshwa's officer, Trimbakji Dengle.

When their power had been ascendant, the Marathas had made sustained efforts to encourage the development of trade and agriculture in many areas that came under their rule (Hardiman 1987a: 34). Kanbi peasants were encouraged to settle in the Rewakantha region through grants of land and initial tax holidays. Almost the entire eastern Mahals (Gujarat) was cleared of forests and brought under the plough by the beginning of the nineteenth century. Under the British, the process of 'settlement' continued apace. Kanbi peasants migrated from Gujarat to Khandesh and took over land (Kulkarni 1983: 266). The Kanbis were looked upon with approval by the British for they were 'model' husbandsmen, innovative and quick to seize the opportunity afforded by the British to 'improve' agriculture, or, in other words, to become commercial farmers.

In order to assure the Kanbis some degree of security, the British had to deal with the vexing problem of the Bhils. According to the Khandesh Gazetteer, the hills were studded with Bhil settlements, 'from a few huts of petty freebooters to grand encampments of powerful chiefs. In the north [of Khandesh], from Kakarmunda to Burhanpur, the Satpudas teemed with the disaffected. . . . The roads were impassable, and in the very heart of the province villages were daily plundered, and cattle and people carried off or murdered. So utterly unsafe did the husbandsmen feel that they refused seed or tillage advances' (KG 1880: 257).

In 1819, there were widespread Bhil 'disorders' all over Khandesh. In 1822, there was another outbreak in the Satpuras. 'The troops were strengthened, the hills overrun, the Bhils scattered, and their settlements destroyed. For two years these fierce retributions went on. But though many were caught and killed, fresh leaders were never wanting, their scattered followers again drew together, and quiet and order were as far away as ever' (KG 1880: 258). For the rest of the century, peace and order remained remote as the Bhils continued resisting rule by outsiders.

The Policy of 'Pacification'

When military action failed to quell the Bhils, the British decided to try a different tack. They chose to investigate 'Bhil disorders' in order to understand what caused them and, through that knowledge, control them and prevent their recurrence. Thus a policy of 'pacification' came to be implemented. Elphinstone, the Governor of Bombay, determined to try these gentler measures. In 1825 orders were given that fresh efforts be made to encourage the wild tribes to settle as husbandsmen, and to enlist and form a Bhil Corps (KG 1880: 258).

The formation of the Bhil Corps would not have happened without Lieutenant (later, Sir) James Outram, whose 'skill and daring as a tiger-hunter, freehanded kindness, and fearless trust in his followers won the Bhils' hearts' (KG 1880: 258). British accounts of Outram's exploits among the Bhils in the hills are eloquently romantic:

Indulging the wild men with feasts and entertainments, and delighting all with his matchless urbanity, Captain Outram at length contrived to draw over to the cause nine recruits, one of whom was a notorious plunderer and had, a short time before, successfully robbed the officer commanding the detachment which had been sent against him. This infant corps soon became strongly attached to the person of their new chief, and entirely devoted to his wishes. Their goodwill had been won by his kind and conciliatory manners; while their admiration and respect had been thoroughly roused and excited by his prowess and valour in the chase (Graham, quoted in Varma 1978: 10).

The British policy of pacification was part of a larger dynamic aimed at establishing a stable colonial government which would oversee extraction. The entire region was divided into Bhil agencies. An administrator was appointed to oversee each agency, his word backed by the coercive strength of the Bhil Corps.

The duties of the Agents were heavy and varied. Gangs still in revolt had to be reduced and order kept, offenders punished or committed for trial, disputes settled and complaints redressed, and pensions paid and people led to settle to steady work. As far as possible, registers of the different tribes were kept; the chiefs won by rewards and pensions, their hereditary claims to guard the passes were carefully respected, and tillage was fostered by grants of land, seed and cattle (KG 1880: 258).

The Bhils of the plains who surrendered to the British administration were the object of much approval:

[The Bhil] feels a relish for that industry which renders subsistence secure, and life peaceful and happy. He unites with the ryot in the

cultivation of those fields which he once ravaged and laid waste; and
protects the village, the traveller, and the property of Government, which
were formerly the objects of his spoliation. The extensive wilds, which
heretofore afforded him cover during his bloody expeditions, are now
smiling with fruitful crops, and population, industry, and opulence are
progressing throughout the land. Schools have been introduced for the
benefit of the rising generations; and the present youth, inured to labour,
and sobered by instruction, have lost the recollection of the state of
older times when, from their insular position, the tribe retorted ven-
geance and hatred upon their oppressors (Graham, quoted in Varma
1978: 11).

Of course, the British did not include *themselves* in the category
of oppressors; instead, they congratulated each other for bringing
civilization to the Bhils. In the case of Alirajpur too, the British
Resident at Indore wrote to his superior expressing his satisfaction
at 'the advantages of the appointment of Bheel [sic] Agents. . . . '
(NAI 1845).[5] The mission of subduing the Bhils and reclaiming
them from savagery necessitated the paradox of 'pacification'
through violence. Outram's Bhil Corps was deployed to root out
rebellion in the hills, the remaining stronghold of the Bhils. In
1830, the Corps marched on the Dangs and subdued the chiefs.
In 1831, they were sent against the Tadvi Bhils of Adavad in
north-east Khandesh. In 1841, Ahmednagar Bhils plundered the
government treasury at Pimpalner. In the same year, Bhamnia
Naik broke into rebellion and attacked a village in Sultanpur. In
1842, Tadvi Bhils plundered Savda and Yaval. The chief of Chikli,
Kumar Jiva Vasava, defied British authority in 1846 and took to
the hills. Troops, including the Bhil Corps, were sent against him
and he was captured only after a bloody battle. The frequency
and violence of these skirmishes challenges optimistic accounts of
that period with their claims of British rule ushering in peace,
progress, 'industry and opulence'. The ruthlessness of the military
operations made clear that pacification was only a more organized
form of coercion.

---

[5] *Administration Report of Ali Rajpore*, 21 February 1846. From Hamilton,
Resident at Indore, to H. Currie, Secretary to the Govt of India with the
Governor General. Nos. 180–1 F.C., NAI.

Reasons for Resistance

Despite British co-optation of some militant Bhils into an indigenous force like the Bhil Corps, and their deployment against
other Bhils, resistance continued unabated throughout the region.
Why did the Bhils keep challenging the might of the British armed
forces? How could they repeatedly withstand the onslaught of the
Bhil Corps, the Nimar International Police, and other military
and paramilitary forces?

According to Ranajit Guha

there is hardly any instance of the peasantry, whether the cautious and
earthy villagers of the plains or the supposedly more volatile adivasis of
the upland tracts, stumbling or drifting into rebellion. They had far too
much at stake and would not launch into it except as a deliberate, even
if desperate, way out of an intolerable condition of existence. Insurgency,
in other words, was a motivated and conscious undertaking on the part
of the rural masses. Yet this consciousness seems to have received little
notice in the literature on the subject (R. Guha 1988: 46).

Indeed, historiography about rebellion mainly consists of two sorts of explanations. Resistance is seen as either a spontaneous response, 'exemplified in those periodical outbursts of crime and lawlessness to which all wild tribes are subject' (R. Guha 1988: 46), or as a reflex action prompted by being driven over the edge by economic and political deprivation. British administrators also resorted to such explanations for Bhil resistance. Thus the super-intendent of Alirajpur referred to 'the ignorant and uncivilized Bheels [sic] of this state . . . [whose condition] is too well known to be described here. Because on simplest provocation and trifling causes they kill each other' (NAI 1896).[6]

In the other sort of explanation, rebellion was reductively attri-buted to economic and political factors. Economically, things *were* a lot worse off for the Bhils. The British had tightened the system of taxation, both where they ruled directly, such as Khandesh, and where they managed by proxy through the Rajput princes, as in Alirajpur. The limited bureaucratization that had occurred in the princely states had been further elaborated. The administration was extended into specialized departments in charge of forest, excise, public works, police and health (Aurora 1972: 16).

With the expansion of the state, the figure of the *vania* (trader-moneylender) became ever-present in tribal life, both as financier of agricultural operations, and as local agent for the collection of land tax (Hardiman 1987a: 28).[7] The vania was best situated to do this, for his dealings with the Bhils had established a relation-ship which the state could use to extract a surplus. The first trade links between the vania and Bhils were those of barter for items such as salt, iron and cloth. A demand for credit emerged as the need for goods did not always coincide with the annual agricul-tural cycle. Once in debt, a cultivator would have to sell his produce to the vania who would chalk it up to the cultivator's account, but the crippling rate of interest ensured that debt was perpetuated as long as possible.

---

[6] *Rules for Extradition of Criminals between Chota-Udaipur and Ali Rajpur*, 7 July 1896. From the Superintendent, Alirajpur, to the Political Agent, Bhopawar. Bhopawar Political Agency. S. No. 5. 1/67/1896. NAI.

[7] So pervasive was the vania that he is incorporated in the sacred wall paintings, *Pithora*, of the Rathva Bhilalas of Chhota Udaipur. He is depicted inside a horse carriage and he is called *Valyo Vanio*, 'the comeback *vania*', for 'he piles up so much interest that the "returning" never ends' (Jain 1984: 36).

The vania would advance seeds to the cultivator and lay claim to the greater part of his produce, denying him free access to the market. The vania thus took full advantage of fluctuations in price, advancing produce in 'loans' when prices ruled high and taking 'repayment' when prices were low (Hardiman 1987a: 11–13). This relationship between the Bhils and the moneylender was an available conduit which rulers could use to their advantage for collecting taxes. In many cases, the vania would take over produce from the Bhils and, instead of paying them in cash or in kind, would directly pay taxes on their behalf.

The poverty of Alirajpur meant that every new exaction brought people closer to the brink of social disaster. The wringing of 'surplus' from small cultivators in the hills was difficult in the best of times for they had so little to surrender; it became intolerable in years of drought. In 1845, for instance, the British Resident summed up the tax collection situation after two successive dry seasons: 'The prospects for the present season are, I regret to say, very disheartening, owing to the want of rain, the crops have failed below the Ghauts, and the price of grain has so risen as to occasion great anxiety for the consequences. No endeavours shall be spared to maintain tranquillity; but it is to be apprehended that robberies will be frequent . . .' (NAI 1845).[8]

In the same year of drought, a somewhat sympathetic Bheel Agent wrote to his superior:

From the high rate at which grain of all kinds is selling in the whole of the districts below the Ghauts, I am apprehensive that cases of felony this year will be on the increase; but my best exertions will be used to induce the [moneylenders] to support their cultivators until the present difficulty is over. Without their assistance it is impossible for a Bheel to support himself and his family without having recourse to plundering . . . for although greatly addicted to intoxication the crimes arising from it are quite of a different complexion, such as abduction, seduction and drunken affrays . . . I consider that if it were not for this degrading vice, a more harmless race of men when kindly treated do not exist, and that it is only when driven to desperation by the oppressive and tyrannical

---

[8] *Administration Report of Ali Rajpore*, 21 February 1846. From Hamilton, Resident at Indore, to H. Currie, Secretary to the Government of India with the Governor General. Nos. 180–1. F.C. NAI.

measures of their rulers that they are, generally speaking, led to commit excesses . . . (NAI 1845).[9]

Despite partial remission of revenue, despite the widespread deployment of troops 'to maintain tranquillity', displaying constancy in the face of kindness or cruelty, the Bhils still rose in rebellion, swooping down from the hills to carry away grain and cattle, retreating into their last remaining stronghold, the forest. This time — the middle of the nineteenth century — was marked by many insurrections all over the land, culminating in the events of 1857.

1857 and the Bhils

During these troubles considerable alarm was felt by the approach, to the very borders of Khandesh, of the rebel troops under the Maratha leader Tatya Tope (KG 1880: 263). In November 1858 Tatya crossed the Narmada and marched west, passing within thirty miles of Burhanpur. His troops plundered a village close to Sendhwa, then went on to rob the post and destroy the telegraph wire on the Bombay–Agra road. After that they apparently changed their plans and, crossing the Narmada, went towards Gujarat. They were met and routed by British troops at Chhota Udaipur. It was then feared that they would re-cross the Narmada and attempt to enter Khandesh through Akrani. Troops were sent to Sultanpur and Taloda, but the alarm subsided as it became known that the rebels, baffled by their attempt to re-cross the Narmada, were moving east towards Khandwa (KG 1880: 263).

Tatya Tope's frustrated foray marked the end of the Mutiny. By 1859 the British had squashed the widespread uprisings in north India, and Lakshmibai of Jhansi had been killed. But even when the Mutiny had been suppressed all over India, the Bhils fought on. In May 1860 the Bhils, under Nimlia Rutnia and other Badwani Naiks, descended from the hills to raid the plains. Later in the same month, they plundered a village in Sultanpur in Khandesh. In June, Bheema Naik's band attacked the Badwani village of Pansemal. The Ubaidaghar Bhils threatened to burn the

[9] *Administration Report of Ali Rajpore,* 21 February 1846. From Capt. D. Wilkie, Bheel Agent, to R.W. Hamilton, Resident at Indore. Nos. 180–1. F.C. NAI.

lines at Badwani. The most famous rebel among the Bhils, Khajya Naik, was at large, his men intercepting and looting travellers throughout the Ghats. Early in June, with a force of about 750 Bhils, Khajya seized twelve camels laden with treasure from the Bombay–Agra road. The British started in pursuit but found that 'the Naik had made good his entry into the difficult fastnesses in the Satpuras' in the Badwani territory. According to the Khandesh Gazetteer, Khajya held out for several months and, along with his army of Bhils aided by Arab and Makrani mercenaries, fought the British at Ambapani (KG 1880: 262). Manohar records that there were 3000 rebels at Ambapani where, after a fierce battle, 460 were captured, but Khajya managed to escape. An alarmed Lieutenant Dysart, Deputy Bheel Agent and Political Agent, shaken by the violence of the widespread rebellion, wrote to Colonel Sir Shakespear, Agent Governor General for Central India, that 'the rising is general' (NAI 1860).[10]

Resistance and Rebel Consciousness

While noting the simultaneity of these events, I am not suggesting orchestration — that the Bhils had a co-ordinated goal which they achieved through co-operative action. While we know something about the increased economic exploitation of the Bhils, we know virtually nothing of the way in which *they* perceived that relationship. What were they fighting for? To what extent could the many points of revolt be joined together into a common picture of anti-colonial struggle? This issue of insurgent consciousness bears upon the general theme of separating people's own understanding of their actions from the claims made about them by others. By highlighting the problematic, often contradictory, nature of the consciousness of dominated people, we approach a more realistic and sympathetic understanding of the possibilities and limitations of resistance.

Most Bhil insurgency consisted of looting and plunder of non-adivasi villages in the plains. Such an idiom of protest seems to point to destitution and the demonstration effect as sources of

[10] *Insurrection of the Bheels of the Satpura Range in Burwanee and Khandeish. Removal of Lt. Dysart from the Appointment of Deputy Bheel Agent and Political Agent.* December 1860. From Col. Sir R.C. Shakespear, Agent Governor General for Central India, to Cecil Beadon, Secretary to Government of India. Foreign Department. Political A. Nos. 184–230. NAI.

discontent. The Bhils had been pushed from the agriculturally more productive plains to the poorer hills. From that position, the non-adivasi plains below were plump and prosperous targets. A difference between the people of the hills and the people of the plains had already emerged and was to become even more marked after the rebellions. Noticeably, resistance was not directed against the British, but against rich villages which could be attacked more easily and profitably. Also, as illustrated by the example of Khajya Naik, the adivasis were willing to compromise with the British and lay down arms if the terms met with their satisfaction. From 1831 to 1851 Khajya had been in the employ of the British, guarding the Sendhwa Ghat pass on the Bombay–Agra road. In 1857, with 4000 followers, Khajya joined Tatya Tope (WNG 1970: 63). The British sent troops against Khajya, who had to surrender. Khajya was pardoned and restored to his situation, and even collected intelligence for the British as late as the uprisings of May 1860. However, soon afterwards, Khajya claimed 'unadjusted dues' from the magistrate of Khandesh and from the Holkar court in Indore. He then started raiding travelling parties in the Ghats. Khajya's grievances, and those of other chieftains, were couched as complaints to the British about rights denied and privileges withheld.[11] The Bhils were usually willing to reach a negotiated settlement. This disposition indicates an acknowledgement and acceptance, however grudging, of British power, implying that Bhil rebellion should perhaps be viewed as an attempt to wrest concessions from the British; it was an attempt which did not challenge their overall domination.

Driven by colonialism to attack those slightly better off than themselves, could Bhil resistance be interpreted as an implicit critique of British domination? This seems dubious. The possibilities and forms of Bhil resistance were contingent upon the relations and conditions of power. While the Bhils must have drawn upon their material and ideological resources — bows and arrows, a knowledge of the hills and of the opportune moment, a sense of injustice — selecting and transforming them to cope with new conditions of subordination, they were also *limited* by those very resources and by those conditions of subordination. The constrained character of their opposition makes it

[11] Ibid.

presumptuous to speak of an insurgent consciousness. Bhil strategies of resistance must be understood in terms of the limitations placed by the changing contingencies of power.[12]

Mathvad and Alirajpur under the British

Conditions of rebellion were somewhat different in the states of Alirajpur and Mathvad. These states, like others in central India, remained nominally independent from the British and under princely rule, either because the rulers were former allies or because their states were not worth the bother of a battle. These states, though on paper accorded territorial integrity, were in practice part of British India. A Political Officer was posted to every state, to advise the ruler as well as to keep a watchful eye on him. If the policies of a prince were lunatic or tyrannical or both, which was not infrequently the case, he was asked or made to abdicate, and replaced by another prince of the same blood, chosen by the British (Moraes 1983: 19). Lunacy was sometimes an arbitrary label, an excuse to remove a recalcitrant ruler. Nonetheless, this rule by proxy cushioned to some extent the populations of princely states from the thoroughgoing changes that British India underwent.

How much of this history applies to the village of Anjanvara in the thakurat (minor estate) of Mathvad? As late as in 1846, the area was described as unexplored, 'the country one entire jungle with immense tree forest interspersed with ranges of low barren hills until they approach the Nerbudda where they rise to a great height. There are few Bheel villages to be met with among them' (NAI 1845).[13] An account from 1866 states that 'the Mutwarh Chiefship is a very petty one, its annual income being about Rupees 1000. The country is very wild and entirely populated by

---

[12] Which is not to swing to the extreme position that E.J. Hobsbawm adopts in his book *Primitive Rebels*, calling 'social bandits' and others 'pre-political people' (Hobsbawm 1959: 2). Even though 'the political allegiance and character of their movements is often undetermined, ambiguous or even ostensibly conservative', their politics cannot be dismissed; it must be accepted for its meaning and intrinsic worth in *their* lives.

[13] *Administration Report of Ali Rajpore.* 21 February 1846. From Capt. Wilkie, Bheel Agent, to R.W. Hamilton, Resident at Indore. Nos. 180–1. F.C. NAI.

a very turbulent race of Bheels, who bordering on the Rewa Kanta
Agency [Gujarat] are continually making marauding expeditions
into it' (NAI 1866).[14] In 1865, Major Cummings wrote that 'the
protection of Allee Rajpoor will be advantageous to Mutwarh
which has a wild and turbulent population of Bheels and Bhillalas
over whom, owing to the remote position of the district, the Bheel
Agent can exercise but little influence . . . ' (NAI 1866).[15]

Mathvad was also part of the widespread unrest of 1857. Ram
Singh, the Thakur of Mathvad, aided the rebellion and spent five
years in jail in Mandleshwar as a consequence. According to the
Bheel Agent's Report of 1862, 'the Bheels, of which [Mathvad's]
population is chiefly composed, are stated, like those in Chiculda
and Burwanee to have given much trouble since 1857, and to
have greatly injured the Revenue of the State by their excesses'
(NAI 1862).[16] It was decided that Mathvad needed the stricter
supervision of the British, and investigations into conditions there
were started forthwith. The improving gaze of the British landed
on revenue collection. The Political Agent, Capt. Cadell wrote of
Mathvad: 'The Land Revenue is collected at rates ranging from
Rs 0–13–6 to Rs 13–8–0 per plough according to the ability of
the cultivator to pay — not according to the quantity or quality
of land held by him' (NAI 1869).[17] '[T]he result of the inquiry
into the affairs of Mutwarh . . . has been to raise the land revenue
nearly 50 per cent . . . ' (NAI 1870).[18]

Despite continued Bhil resistance, British rule became more
consolidated after the insurrections of 1857. Alirajpur underwent
a change of rule in 1869 when the Political Agent sent a vivid and

[14] From Capt. Bannerman, Assistant Agent Governor General, to Lt. Col.
R.J. Meade, Agent for the Governor General in Central India. Bhopawar
Political Agency. S. No. 17. 1/153/1870. NAI.

[15] From Major W.G. Cummings, Bheel Agent, to Lt. Col. R.J. Meade,
Agent for the Governor General for Central India. Bhopawar Political Agency.
S. No. 17. 1/153/1870. NAI.

[16] *Report on the Districts and Petty States under the Bheel Agency for the year
1861.* By Major Cummings, Bheel Agent. Foreign Department. Political A.
1862. Nos. 156/158. NAI.

[17] From Captain Cadell, Political Agent Bhopawar, to Col. Daly, Officiating
Agent, Governor General for Central India. Bhopawar Political Agency.
S. No.17. 1/153/1870. NAI.

[18] From Cadell to Capt. Berkeley, Officiating First Assistant to the Governor
General. Bhopawar Political Agency. S. No. 17. 1/153/1870. NAI.

damning report of its ruler's incompetence to his superior, justifying further British intervention in Alirajpur's affairs:

The Maharaja Gungadeo is more than ever addicted to the immoderate use of opium and spirits, and the epileptic fits to which he is subject, brought on as they are entirely by his own vices, are more frequent and of a severer type . . . [H]is appearance when we met was most deplorable. He could hardly crawl the few steps from the entrance of my tent to his chair when he came to visit me . . . The Chief leaves everything in the hands of his Kamdar Cagee Abdul Rouf . . . who is very indifferently educated, utterly unacquainted with the simplest Revenue rules or questions, has no knowledge whatever of judicial procedure — arrogant and insulting in his manner, and very grasping and avaricious . . . the lawlessness and turbulence of the wild tribes inhabiting that part of the country make such a state of things very serious (NAI 1869).[19]

With this battery of charges, the British brought down Rana Ganga Dev and placed the state 'under superintendence'. Roop Dev, the Chief's younger brother was 'given a place in the administration' (Luard 1908a: 599). While British supremacy had manifested itself as early as 1818, direct control was now established even more firmly.

Changes in Land Use and Tenure

Around this time, adivasis were 'encouraged' to stop *dahia* (shifting cultivation) and settle down. Guha and Gadgil observe that

almost without exception, colonial administrators viewed [shifting cultivation] with disfavour as a primitive and unremunerative form of agriculture in comparison with plough cultivation. Influenced both by the agricultural revolution in Europe and the revenue-generating possibilities of intensive (as opposed to extensive) forms of cultivation, official hostility to [shifting cultivation] gained an added impetus with the commercialization of the forest. Like their counterparts in other parts of the globe, British foresters held [shifting cultivation] to be the most destructive of all practices for the forest, not the least because it competed with timber operations (Guha and Gadgil 1989: 152).

[19] *Reports of Captains Bannerman and Cadell on the Maladministration of Ali Rajpoor State, Deposition of Rana Gungadeoji, and Placing the Government under British Management.* Bhopawar Political Agency. 1/151/1869. NAI.

Earlier there had been no clear demarcation between revenue and forest land; administration of both was vested in the same official. Within the jurisdiction of his village, the headman gave permission for expanding cultivation. With the tightening of colonial rule, the authority of the headman to permit fresh clearings in the forest came to be abrogated and was instead vested in state officials (Nath 1960: 27). In the process, property rights were sharply redefined. However, in the more remote parts of Mathvad, such as Anjanvara, customary rights continued to be respected. The ruggedness of the terrain and the absence of roads limited the extent to which forests could be exploited for timber. As long as the state received its due, the thakur turned a blind eye to nevad, the expansion of cultivation. Without competition from large scale state-sponsored deforestation and the pressure of increasing population density, nevad could remain politically and ecologically tolerable. It must be emphasized, though, that Mathvad was unusual in its geographic isolation. For almost a hundred years it remained insulated from the widespread commodification of the forest that occurred around this time.

From being an intrinsic part of peasant agriculture, forests came to be inserted into a commercial economy which sharply undermined the ecological basis of subsistence agriculture, hunting and gathering. Adivasis were being increasingly excluded from the forest and their customary use rights restricted. Yet in most places this was not justified by the British on grounds of environmental protection, for no policy of conservation was instituted. Land was leased to contractors whose activities turned vast tracts of forest into semi-barren land (Aurora 1972: 87). Indian teak was extensively used in ship-building for the royal navy in the Anglo–French wars of the early nineteenth century and by merchant ships in the later period of maritime expansion, as a substitute for the depleted oak forests of England and Ireland (Guha and Gadgil 1989: 145). The expansion of the railways (1870–1910) also resulted in the widespread destruction of forests. The British 'reservation' of the forests resulted in large scale felling in this region too. In the summer of 1877, 'though the [Narmada] was unusually low, a flotilla of 625 logs and 6000 rafters was . . . floated from the north-east of Akrani to Broach, where it fetched more than three times the amount spent on felling, dragging, and floating it down' (KG 1880: 9).

The settling of shifting cultivators was simultaneous with the settling of non-adivasis on adivasi land in the plains. According to Manohar, after 1850 the population of Khandesh increased steadily. Kanbi-Patidars from Gujarat occupied land around Shahada and Taloda, for the British wanted 'progressive' farmers who would cultivate cotton and opium. Adivasis were pushed off the land and made to serve as indentured labour. In Gujarat, too, land passed into the hands of Patidar settlers (Epstein 1988: 22). Hardiman says that Leva Kanbis [Patidars], who farmed near Jhalod, were moneylenders and advanced money to Bhils with an eye to gaining ownership of their land in the settlement of debts. 'Though only a few, their activities aroused considerable resentment among the Bhils' (Hardiman 1987a: 17).

The Origins of Anjanvara

According to people in Anjanvara, their ancestors lived in Gujarat, where they fought with the Patidars for whom they laboured. When escaping, some fled on horseback and some on foot. Of these, some fell off and were left behind by the others. This was the beginning of the Padyarka clan to which the men of Anjanvara belong; their name deriving from *pad*, the verb for 'to fall'. While this account closely mirrors other historical sources, some narratives seem to blend history and myth in ways which emphasize magical origins. One story tells how there was only one clan in the beginning, which made the rule of exogamy impossible to maintain. In order to avoid incestuous unions, this clan then split into two — Padyarka and Bugwadia.

Another story attributes the settling of Anjanvara to the magical powers of two ancestors, Raisingh and Malsingh, who drank a pot of *huru* (liquor) each and went to the market in Mathvad. A trader woman sat selling her wares. Raisingh and Malsingh held on to her breasts and refused to let go. A crowd collected and rained blows on them, but they held on. The market erupted in confusion; the king came to see what the noise was about but Raisingh and Malsingh fought him too. Frightened, the king ran off to his manor, shutting the gates behind him. But Raisingh and Malsingh came after him. They tore down the gates and, charging in, began shaking the pillars of the manor. The king came out in abject terror and begged for mercy. He said: 'I will give you all you want, only leave me alone.' The king offered them gold and silver, and all the most precious things in his possession, but they were steadfast in their refusal. Finally he said, 'I will give you Anjanvara,' and only then were Raisingh and Malsingh appeased. And so the village came to be.

While the people of Anjanvara trace their ancestry to Gujarat and the experience of being dispossessed by non-adivasis there, the process of land alienation seems, for the most part, to have missed Alirajpur. Some land *did* pass into the hands of moneylenders because of peasant indebtedness, but its extent was probably insignificant. According to Aurora, most of the land in Alirajpur was of sub-marginal productivity; vanias were not interested in owning land in the region, preferring to leave cultivation in the hands of adivasis whom they could more profitably exploit through unequal terms of trade (Aurora 1972: 181). By this time, adivasis had been pushed into the hills. Their holdings were unattractive enough to remain uncoveted.

The Commercialization of Natural Resources

The settlement of non-adivasis in the more fertile plains was related to the massive increase in the commercialization of agriculture for imperial sourcing. Seed was imported from the USA, Egypt and France to improve the Indian varieties of cotton. Cotton cultivation which had existed in this region since the beginning of the nineteenth century, was later also encouraged to create an alternative to American sources blocked by the Civil War during the 1860s (KG 1880: 155ff). By 1878 the area under cotton in Khandesh had outstripped the tillage for *juvar* (sorghum), the staple food crop (KG 1880: 149). In Gujarat, cotton and tobacco were the cash crops upon which production was concentrated.

As the area under food crops declined, the trade in food grains rose. In the entire region, bullock carts laden with produce travelled by road to distant urban centres in Gujarat and Bombay; during the 1890s the building of a railway line connected Dahod, a town close to Alirajpur, even more securely to the cities. Hardiman quotes a Deputy Collector of the time: 'The inhabitants of the Mahals [the area north-west of Alirajpur] have begun to feel the advantages of the introduction of the railways to their very threshold . . . ' The quotation is ironic for, says Hardiman 'the benefit to the Bhil cultivators would have been minimal as they never received the market value of their crops. It was, rather, the grain dealers who were enriched by the opening of new markets through the railway' (Hardiman 1987a: 19).

The hastening of grain movement by the railways did not occur directly in Alirajpur, but the kingdom *was* stirred by the winds of change. As in other parts of India, taxation and land tenure were the two systems of exploitation to undergo the most radical transformation under colonialism (Wolf 1982: 247). The commercialization of food production in Alirajpur was encouraged by shifting to monetary revenue and excise payments. The collection of excise duties was given out on contract to traders who would advance loans to the Bhils in exchange for first rights to their produce (Luard 1908: 610). As mentioned earlier, the vanias, who were intermediaries between the administration and the people, encouraged taxpayers to grow cash crops such as sesame and groundnut which were more remunerative, strengthening their ties with distant markets (Aurora 1972: 88).

As the example of 'the British enquiry into the affairs of Mathvad' shows, colonial efforts to survey and enumerate, ostensibly to aid good governance, invariably ended up to their monetary advantage as well. The administrators made strenuous efforts to know in order to tax and take away; only vociferous objections from the populace held them in check. In Alirajpur, to this end, the Native Superintendent wrote that, ' . . . I am of the opinion that all the Mangoe [sic] and Mowa trees within the Rajpore state should be enumerated and registered in a Book, which will prove of service when some disputes arise about possession. Also, it is possible, when all the trees have been enumerated, that a tax, which will be deemed fit, may be levied' (NAI 1870).[20]

With this view, measures were adopted to ascertain the total number of trees, but the work was stopped when the Bheel Agent was informed of the cultivators' 'displeasure'. The Superintendent went on dismissively, 'On this, I minutely inquired into the matter and informed Captain Cadell that they were not at all displeased on account of the trees, only they had entertained a doubt, which was removed by discontinuing the enumeration measures. But it is usual that the cultivators grumble and murmur at every new arrangement' (NAI 1870).[21] Similarly, a list of about a hundred local trees and their various uses was prepared in 1894 with a view to their commercial exploitation (NAI 1894).[22]

The crushing pincers of the state's revenue exactions and the depletion of natural resources must have made survival even more precarious than usual, necessitating seasonal migration to the fields of Gujarat, Nimar and the Tapti valley in years when famine stalked the land.

Ironically, the only historical evidence of the destruction of

[20] From the Native Superintendent of Ali Rajpore to Col. Blair, Officiating Bheel Agent. 27 August 1870. Bhopawar Political Agency. S. No. 16. 1/152/1870. NAI.

[21] Ibid.

[22] Report from the Ameen Alirajpur. 1894. *Yaad sadar aam riyasat Alirajpur. Babad haal darakht jangal Alirajpur.* NAI. This inventory is also interesting for its attempt at compiling tribal knowledge about 'bio-indicators', i.e. correspondences between the quality of flowering of certain wild tree species and the yields of crops. For example, the flowering of *babul* (*Acacia nilotica*) in profusion augurs an abundant *bajra* (*Pennisetum typhoides*) crop.

forests and its consequences is the immediate and enormous increase registered in the state's income, mainly from land revenue,
excise dues and forests.[23] Statistics for Alirajpur state's receipts
from 1901–2 to 1937–8 show a threefold rise in total revenue
over a period when the rupee did not change significantly in
value.[24] Land revenue grew from Rs 38,000 in 1901–2 to
Rs 241,540 in 1937–8, an increase of 635 per cent. Excise receipts
rose from Rs 16,000 to Rs 37,521, an increase of 234 per cent.
Over the same period, the forest department registered the greatest
rise, mainly from massive deforestation and the sale of timber;
while it had generated Rs 11,000 in 1901–2, in 1937–8 it
produced Rs 113,564 — an increase of 1032 per cent. Expenditure statements show that a good third of this revenue went
towards the maintenance and expansion of police and revenue
collection agencies (Aurora 1972: 82).

The continuing development of the revenue system now linked
the villages even more firmly to the district headquarters. For the
most part, the intermediaries in the system, revenue officials as well
as traders, were all non-adivasis recruited from immigrant communities. Aurora remarks that the British revenue system tried to create
a bureaucratic link between the tribal farmers and non-tribal officials, where 'the impersonality of the bureaucratic set-up is usually
compensated by its rationality. But this set up was only partially
bureaucratic since rules regarding the collection of revenue had not
been rationalized, as a result of which the scope for ruthless exploitation of the tribals was very much extended' (Aurora 1972: 77).

Once again, the collusion of the market and the state, the vania
trader and the *talati* (petty official), must be emphasized. In
expropriating peasant surplus, the vania's interests coincided with
that of the British administration; the courts of British justice were
manipulated by moneylenders to their own ends (Hardiman
1987a: 35).[25]

<hr>

[23] For estimates of the extent of deforestation in colonial times in other
regions of British India, see Gadgil and Guha (1992).

[24] Luard's Alirajpur Gazetteer has revenue and expenditure figures for 1881–
90 and 1891–1900 but, considering that they pertain to ten-year periods, they
are suspiciously low, being even less than the amounts for single-year periods
such as 1901–2. For this reason, comparison is only based on the periods for
which we have uniform data.

[25] The convergence of colonial and mercantile capitalist interests was

The dominance of commerce and colonial administration over the tribal hinterland was spatially represented by the emergence of towns at places where haats (weekly markets) were held. According to traders in Valpur and Chhaktala (towns close to Mathvad), the institution of the haat has existed since known times, but all the present permanent traders are immigrants who settled there four or five generations ago (Aurora 1972: 69). The haat metamorphosed into small towns which were locations of institutions of political authority as well as points in the trade of produce. As centralized points of control and collection, these towns served as nodes which channelled regional surplus into a national market. Forest produce such as timber, lac, *mahua* (*Madhuca indica*) flowers, for example, would go from the haat 'to Dohad (Panch Mahals) and Kukshi (Dhar), the nearest foreign markets' (Luard 1908a: 604). As primary sites where subordinated tribal communities interacted with non-tribals, towns served also as centres of cultural exchange and differentiation, where the ways of adivasi and non-adivasi living were counterposed.[26]

The development and consolidation of systems of exploitation by a colonial bureaucracy in alliance with various indigenous classes did not go unchallenged; through popular revolts the adivasis reiterated their opposition to *begaar* (the practice of forced, unpaid labour for the state) and to the exactions of vanias and corrupt government officials. Oppressed by the demands of the state, people led precarious lives. Any misfortune could bring forth disproportionate misery and violent reaction. Unrest always reared its head in years of drought when the state's demands for 'surplus' from a starving population became more outrageous than usual. A multitude of often disparate grievances would then coalesce to forge a common cause against the state.

Popular Resistance and the State

The year 1881 was one such time of rebellion. The state of

---

particular to this set of extractive conditions; in other parts of India, British policy addressed the issue of agricultural indebtedness more critically, identifying it as an obstacle to the emergence of capitalist agriculture. [See D. Ludden (1984: 69–70) for a partial survey of literature on agricultural credit in India.]

[26] The issue of exchange and differentiation between tribal culture and 'the broader Indian civilization' is discussed in more detail in Chapter 4.

Alirajpur was rocked by a dispute over succession because the king, Roop Dev, died childless and the British decided to appoint Bijai Singh from the Sondwa Thakur's family to the throne. This was not approved of by Jit Singh, the Thakur of Phulmal, and the Makranis (mercenaries from Afghanistan who had settled in the region), whose power had been eclipsed by the British. They united with

the Bhils who were in a distracted state, as owing to want of proper supervision, the *patwaris* and district officials had extorted considerable sums from these people by raising the assessment as high as they liked. Joining with Chhitu Bhil, Patel of Sorwa, and Bhawan Tarvi of Tokria-Jhiran [Jit Singh] collected the discontented faction and plundered villages of Nanpur, Chhaktala, and Bhabra, while even Rajpur was threatened (Luard 1908a: 599).

Dad Muhammad, the leader of the Makranis, summoned men from the states of Khandesh, Chhota Udaipur and Gujarat to aid his cause. The British moved with 36 Lancers of the Central India Horse and 63 men of the Malwa Bhil Corps; the rebellion was suppressed after a battle at Sorwa Pass.[27]

Aurora supplements this account in the Alirajpur Gazetteer with his own research:

According to local tradition, the years of rebellion, 1881–82, were particularly bad for the farmers and famine was rampant throughout the State. But in the commercial villages and towns most of the traders had large stocks of hoarded cereals. According to one very old informant, when the rebels reached the outskirts of Rajpur, Ganga Deo and Rup Dev's mother (the queen mother) went with a plateful of money to Chittu Patel and asked him to spare her *Ryayah* (subjects) and take away the money. Chittu refused the money and told her that he had nothing against the Raj (crown), he was not a *lutero* (robber), he only wanted that the *vanyas* should open up their grain stores so that his hungry followers could have something to eat (Aurora 1972: 77).

This episode goes to the heart of the complex relationship between adivasis and the state. In the person of the king were fused divine sanction and temporal power; kingship was celebrated in the festival of Dussehra at Alirajpur and Mathvad. The king was worshipped as 'protector, arbiter, ritualist and embodiment

---

[27] Ironically, Chhitu Bhil's fortified house in Sorwa, which was captured by the British, is today a police station.

of wealth' (Ludden 1984: 59); his deeds garnered wellbeing for his people. Yet the majestic paternalism invoked and accepted in this relationship did not extend to include the king's men or those responsible for realizing the state's due, and people often chafed under the oppressive yoke of taxation and the requisition of unpaid labour. While resentment was directed against a particular ruler or his laws, the hegemony of kingship remained unchallenged in the ensuing protest. As Scott observes, the key reciprocal duty of elites is to guarantee subsistence (Scott 1976: 188). When even that bargain, forged of the iron of inequality, is broken, a sense of injustice cannot be denied. The famine year of 1881 was such a time when people acted decisively against a moral economy betrayed. It must be stressed, though, that the divinity of royal power was an aura that clothed a changed king — one who continued to derive legitimacy from tradition but who was essentially in thrall of the British. The charismatic authority of the king maintained the mannerisms of princely rule but concealed the colonial powers behind the throne. The potency of this paternalistic idiom has persisted over time, passing down to the modern state, even though the character of the state has shifted.

Matters were more or less 'settled' since then, the 'peace' broken only by famine like the devastating one of 1899–1901 when, in two consecutive years, the annual rainfall averaged 15 inches — much below the normal average of 35 inches per annum (Luard 1908b: 612). This time the British dealt differently with the problem; grain was imported into Alirajpur. Poor houses and famine relief works were started and remissions of revenue were granted. Incidents of looting were sporadic. By this time, life for most adivasis had turned inwards; they concentrated on trying to survive intact — holding their own in the hills.

Nationalism and the Independent State

The nationalist struggle for independence, which mobilized people all over the country in the first half of this century, received surprisingly little support from the adivasis. The Civil Disobedience Movement was marked by widespread defiance of forest regulations elsewhere in India, but no such protests are known to have occurred here (Sarkar 1980). For the Bhils of west Khandesh, people who had fought the British every inch of the way,

nationalism was an ideology emanating from groups whose interests were antithetical to those of the Bhils. The call for non-cooperation and satyagraha was raised by rich Patidar farmers and
moneylenders, their traditional exploiters. While non-cooperation
was eagerly seized upon by the 'progressive' farmers and traders
of prosperous districts in Gujarat, 'only in the more backward
Panch Mahals had the movement failed to take hold . . . Indeed,
in a typical eastern Mahal like Dohad, by 1921 the Bhils who
formed the bulk of the cultivating population had come out firmly
in opposition to the local banias [*vanias*] who favoured non-
cooperation' (Epstein 1988: 43). In Alirajpur, it would appear
that the call for non-cooperation went unheeded for other reasons.
According to Aurora, 'in 1942, when all of British India was
rocked by the struggles of the nationalists, only a few people in
Alirajpur knew about it. The tribals were not even remotely aware
of the nationalist movement' (Aurora 1972: 210).

This apathy towards nationalism is puzzling. Whereas adivasis
seized the opportunities afforded at the time of the 1857 Mutiny
to further their own struggles, they seem to have remained indif-
ferent to the call of the nationalists. While this may be interpreted
as their recognition of the class character of nationalism, and a
refusal to be assimilated into its universalizing discourse, it also
represents a failure on the part of adivasis to appropriate the
ideology of nationalism to their own ends — an omission that
appears curious when we consider the many creative local trans-
formations of the nationalist enterprise that occurred elsewhere in
the region (cf Hardiman 1987b). Alirajpur also remained untouched by the 1939–49 movements against *veth-begaar* (forced
labour)[28] and against the payment of rent to jagirs (feudal estates)

---

[28] The strategy of the campaign against veth-begaar was little short of
brilliant. Mama Baleshwar had written copiously protesting against this practice
of the princely rulers, but had made little headway. Then Mamaji went to meet
the Shankaracharya of Puri and told him that while the upper castes — the
brahmins, kshatriyas and vaishyas — were excused from the requirement of
performing free labour for the king, the adivasis were suffering under its yoke.
Mamaji told the Shankaracharya that adivasis were converting to Christianity
in large numbers in order to escape the king's *veth*, for the British had made
Christians exempt from forced labour. Mamaji said that, if the Shankaracharya
would allow the adivasis to wear the sacred thread of the Hindus, conversions
to Christianity would cease and so would the suffering of the adivasis. After
much thought, the Shankaracharya agreed to Mamaji's suggestion but on the

started by Mama Baleshwar of Bamnia which affected at least nine princely states in present-day north Jhabua and Rajasthan.

After 1947, when India became free, the kingdom of Alirajpur merged with independent India as a constituent of the state of Madhya Bharat. Jhabua district was constituted in 1949 and Alirajpur became a tehsil within it. The services of local officials were incorporated in the Madhya Bharat Administrative Services. Almost the first act of the new regime was to order a detailed land survey for permanent settlement. Titles to land were issued, soil types were assessed, and regular rates of land revenue were fixed. The number of officials in the revenue department was considerably increased. Ironically, land disputes became much more common after the survey. The manipulation of land records by the patvaris became a source of corruption and power in the hands of revenue officials (Aurora 1972: 202).

With the passing of power into the hands of Congress nationalists after Independence, the state embarked upon its project of 'development'. The independent state's agenda of progress called for accelerated industrialization and modernization, together with a populist 'welfare' component in the form of anti-poverty programmes. The state was charged by the Constitution to 'promote with special care the educational and economic interests of the weaker sections of the people, and in particular, of the Scheduled Castes and the Scheduled Tribes, and [to] protect them from social injustice and all forms of exploitation' (GOI 1978: 4).

To this end, Jhabua district and parts of the neighbouring districts of Khargone and Dhar, were declared Scheduled Areas in 1950. Their populations of Bhils and Bhilalas were classified as Scheduled Tribes by 1956 (NCAER 1963: 191). Welfare measures such as ashrams (residential schools), income generation schemes

---

condition that adivasis give up liquor and meat. A jubilant Mamaji widely publicized the Shankaracharya's order and thousands of adivasis flocked to Bamnia for mass ceremonies to don the sacred thread and defy the king's law. The princely rulers retaliated by getting the police to break off the sacred thread from adivasis found wearing it. This too was publicized by Mamaji and the kings found themselves facing delegations of wrathful Arya Samajis, demanding an end to the sacrilege. Mamaji succeeded in stopping unpaid labour as well as making a severe dent in the liquor trade which was controlled by the princely rulers. (Mama Baleshwar, interviewed by Amit Bhatnagar.)

such as the Integrated Rural Development Programme (IRDP), and government-controlled co-operative societies were started. However, most government programmes consisted of handouts in times of distress, especially during drought, which failed to address the underlying causes of adivasi impoverishment. Tribal rights to the forest remained unrecognized; their continued alienation from the land base upon which they depended for sustenance precluded any opportunity for genuine gains in power and prosperity.

This is not surprising when we examine the overall framework of the national policy for tribal areas and tribal communities in the country. According to the Report of the National Council of Applied Economic Research, the aim of the policy

is to integrate the tribal communities within the body politic of the nation. This is sought to be achieved through raising the standard of living in the tribal areas to that in the rural areas of the country. Since the tribal areas are exceptionally backward and primitive, welfare of the people living therein is sought to be achieved through some measures of protection during the next 20 years. This protection is in respect of seats in Parliament and legislatures, reservation in services of all types and categories, educational benefits and economic protection. The aim of the Government is to level up the tribal communities without unduly damaging their social structure or interfering with the way of their lives (NCAER 1963: vi).

To this lofty end

top priority has been given to a programme of rapid industrialization and extension of means of communication to the most interior regions in the State. Our firm view is that the development of land and agriculture alone will not be adequate for the rehabilitation of the tribal communities. Agricultural land is insufficient and cannot serve the needs of even half the tribal population of the State. Fortunately, the tribal areas of the State are rich in industrial and power potential. There is no reason why in the wider interest of the nation and in the long-term interest of the tribals themselves, industries should not be developed and localized in tribal areas (NCAER 1963: vi).

In the 'wider interests of the nation', the state has exercised its prerogative of claiming eminent domain — the greater good of the people — to pre-empt resources for itself.[29] Yet the pursuit of

[29] For a legal discussion of the principle of eminent domain in the context of India's forest policy, see Chhatrapati Singh 1986.

these policies has brought about rapid exploitation of natural resources in tribal areas, violating the interests of dispossessed adivasis. The acceleration of extraction has been matched by the expansion of administrative control to further restrict adivasi use of the land and the forest. These regulations have become the source of a steady stream of remuneration for petty officials who browbeat adivasis who break the law into the submission of bribes and 'gifts' of omission. Meanwhile the forest department, legendary in its corruption, colludes with timber merchants to decimate the forest.

The relationship of the adivasis to their natural environment is mediated by the state in other political realms too. As mentioned before, land settlement resulted in more disputes which are increasingly settled in court and at additional expense (Aurora 1972: 161). The traditional panchayat system of arbitration, administered by elders within the community, has been ignored and judicial authority transferred to the state. Lawyers and policemen do their best to profit by the enforcement of 'law and order'.

Domination in the Form of Caste Ideology

Processes of incorporation into national and international markets, which gathered momentum in colonial times, have gained strength since independence. But along with, and buttressed by, greater commercialization and interaction with the state apparatus, adivasis have also had to contend with another exchange conducted on unequal terms, viz. Hinduization. The haat and town have been the locus where adivasi/non-adivasi boundaries have become most visible. There is the brutal contempt of the Hindu bazaaria for adivasis because they eat meat and drink alcohol. There is the distance and suspicion, born of experience, of the adivasis for the bazaaria. While towns have been sites for the crystallization of attitudes and identities, they have also propagated the dominant ideology of caste.[30]

William Roseberry theorizes that 'given the nature of hegemonic political communities and the political and economic structures in which they are inserted, most alternative movements . . . take a form (and may involve contents) similar to the dominant

[30] A fuller discussion of this point follows in Chapter 4.

culture' (Roseberry 1989: 228). Attempts to Hinduize by imitating the practices of the ritually purer castes have sporadically occurred in this region among the Bhils. Such efforts represent the contradictory impulses of, on the one hand, affirming the caste system by accepting its ideology, and on the other hand, rejecting the position assigned to adivasis by the higher castes.[31] Most commonly, Hinduization requires the abjuration of adivasi practices such as animal sacrifice, the eating of meat and drinking of liquor, and the adoption of Hindu deities. In the 1950s, Alirajpur had such a *bhagat* movement, started by Harinam Singh, a carpenter, whose disciples when coming into town would wear dhotis in the vania style and smear sandalwood paste on their foreheads (Aurora 1972: 246). This movement, like the few others which have followed, was limited for the most part to adivasis who had had prolonged contact with bazaaria domination, i.e. those who lived in, or close to, the town.

For an adivasi, Hinduization is socially extremely painful for it requires a complete repudiation of Bhil/Bhilala identity and community. Someone who becomes a bhagat cannot eat or marry with non-bhagats, an interdiction amounting to almost total ostracism of family and friends. Such a break is contemplated only when the possibility of gains from upward mobility — getting a government job (or anything facilitated by being accepted into the Hindu world) — far exceeds the loss of identity and security. For these reasons, bhagat movements hold little appeal for most people in the hills. Rather than embrace Hinduism, the people of the hills still run away when a government official walks into the village — a response that is changing with the work of those organizing the adivasis.

The urge to act against the exploitation of adivasis by bazaarias in post-Independence India had earlier been harnessed by way of electoral politics. During the general elections of 1952, candidates of the Socialist Party won against the Congress. The party was supported by adivasis who were mobilized through the extensive grassroots campaign of the renowned Socialist leader Mama Baleshwar of Bamnia. Among other election issues, the Socialists promised people free wood from the forest. By 1956, the Congress

---

[31] Chatterjee (1989) describes a similar predicament for low caste Hindus who strive to improve their status.

Party in the tehsil was fully identified with the traders (Aurora 1972: 211). In the elections of 1962, Socialist MLAs won from Jhabua and Alirajpur constituencies, their campaign based on opposition to begaar and to corruption in the forest, soil conservation, police and revenue departments. Some of the most active Socialist workers were drawn from students of the residential schools, identified by their blue shorts, white shirts and red caps, the last explaining why their movement was called the Lal Topi (red cap) Andolan. The movement ended because the limitations of electoral politics precluded any far-reaching changes from being instituted. Socialist representatives, who were in a minority in the state legislature, were ineffectual in bringing about change. Subsequent mobilization in Alirajpur has been in the form of non-party, non-electoral organizing by the Khedut Mazdoor Chetna Sangath and the Narmada Bachao Andolan. From the point of view of the adivasis, the relationship between the people and the pre-colonial, colonial and nationalist state has scarcely altered — the state dominates and the adivasis are made subordinate. The difference is one of degree. Within that relationship of domination/subordination, there have been times when resistance has been overt, events of insubordination that have been acknowledged in official records. But for the most part, people have had to submit, their resentment summed up in the words of the *budva* of Anjanvara — 'The state is a thief'.

The continuity between different regimes is important to remember for, too often, we tend to see the precolonial era as the golden age of moral economy when adivasis lived well, in harmony with each other and with nature. As Roseberry observes:

While pointing out the importance of the past in the present [we] analyze a relatively unambiguous transition from an ordered past to a disordered present. We instead need to view a movement from a disordered past to a disordered present. With such a starting point we can assess the contradictions inherent in the development of . . . consciousness and appreciate that the past provides experiences that may make the transition seem positive . . . Only then can we see the moral economy as a source of protest and accommodation, despair and hope (Roseberry 1989: 58).

4

Bhilalas: Caste or Tribe?

The history of the Bhils and Bhilalas in western India has been a chronicle of incorporation and resistance. The tribal people of the Narmada valley have experienced the expansion of political control over their lives by the state, greater integration into the market economy, and increased domination by the processes of modernization and Hinduization. These processes have met with resistance, but with varying degrees of success. So today, when we talk about 'tribes', who are we talking about? Have tribes melted into the Hindu embrace? Does the concept of 'adivasi' continue to have an existential reality?

The Politics of Sociological Designation

Indian sociologists have long tried to gauge the fit between the concept of 'tribe' and its applicability in India. According to André Bétéille, they have sought a way out of the muddle by calling them all 'tribes in transition' (Bétéille 1986: 299). This does not settle the issue at all because tribes have always been in transition. Also, the notion of 'transition' is an oversimplification; it assumes a linear, evolutionary scheme of change. As Bétéille points out, 'there is reason to believe that both detribalization and retribalization occurred in the distant as well as the recent past' (Bétéille 1986: 310).

The notion of 'tribe' in the Indian context has been hard to pin down because of the porosity of the boundary between 'tribe' and 'non-tribe', both of which have existed side-by-side for centuries. And, as the ethnographies of Muslim Swat Pathans in Pakistan and Buddhist Sinhalese in Sri Lanka make evident, caste as a structural organization is indissolubly linked with a Pan-Indian civilization (Leach 1960), such that caste-like features are found all over the Indian subcontinent, including among non-Hindu communities.

The political controversy over tribe/caste identities emerged when the process of designating or 'scheduling' tribes in India began during British rule. This process was influenced by the ideologies of the scholars in question. On the one side were official anthropologists, mostly British members of the Indian Civil Service, who argued that the aboriginal tribes had a distinct identity that marked them out from the rest of Indian society.[1] On the other side were nationalist anthropologists who argued that tribal people were just a poorly integrated part of mainstream Hindu society. Thus, G.S. Ghurye awkwardly called them 'backward Hindus'. In order to oppose the British policy of divide and rule, he argued that tribes could not be proved to be India's aborigines (Ghurye 1963). From his nationalist perspective, an emphasis on their indigenous status could only appear politically divisive. 'These points of view, though apparently contradictory, have both been accommodated in the present Constitution which recognizes

[1] For instance, C.E. Luard, in the Alirajpur Gazetteer of 1908, classified Bhils and Bhilalas as animists and not Hindus (Luard 1908: 614).

that tribes are different from castes, but treats tribals, with in-dividual exceptions, as Hindus all the same' (Béteille 1986: 317). Even today, the relationship between tribes and Hinduism is a hotly debated issue, where the different stances taken by scholars tend to reflect their political agenda.[2]

The recognition by some sociologists of the difference between tribes and castes resulted in the construction of an ideal-type where isolation, self-sufficiency and autonomy were deemed to be char-acteristics of the tribal condition (Bose 1971); characteristics that, as history tells us, have taken centuries of battering. So when we analyse the relationship between 'adivasis' and the caste system, we must remember that contemporary identities are devised within a larger system of cultural dominance and subordination, and that they acquire different values according to the changing contingencies of power. In all their manifestations, in particular historical circumstances, adivasi identities have been shaped by their condition of subordination to the Hindu system.

In its deviance from the dominant caste ideology, adivasi cul-ture has used the varied resources of given cultural traditions, selecting, transforming and developing them to cope with new conditions of exploitation but remaining limited by those condi-tions. Chatterjee's observation about similar strategies used by lower caste Hindus is relevant here: '[In this] layering in popular consciousness of material drawn from diverse dominant as well as subordinate traditions, the only principle of unity [has been] the contradictory one of simultaneous acceptance and rejection of domination' (Chatterjee 1989: 189). And, one might add, the appropriation of dominant ideologies for one's own ends. Thus, within the constraints set by Hindu domination, there is wide variation in adivasi experience and action.

[2] The World Bank in 1982 defined what constituted a 'tribal' community in order to formulate special protective policies which would ensure that tribal people are not 'harmed . . . by development projects' (Morse and Berger 1992: 71). Using that definition, the World Bank-appointed Independent Review of the Sardar Sarovar Project designated the Bhils and Bhilalas of the submergence zone as 'tribal' people and argued that the project would be inimical to their interests. Such a classification was immediately and indignantly challenged by supporters of the project who said that the review had misunderstood the place of tribes in modern India and that its analysis reproduced the ideological position of colonial anthropologists. Latter-day versions of the nationalist stand are also found among Hindu fundamentalists who claim the unity of tribe and caste.

Hinduization Among the Adivasis of the Nimar Plains

In Nimar, for instance, Bhils and Bhilalas have defined themselves in response to historical processes of a very different magnitude from the adivasis of Maharashtra and Alirajpur. The fertile plains of Nimar were settled in the last century by Kanbi-Patidars from Gujarat. The emergence of the Patidars as the dominant landowning caste which, with the encouragement of the British, grew cash crops, coincided with the creation of a class of marginal farmers and landless labourers which included Bhils and Bhilalas in its ranks. Living within a system of occupational specialization closely related to caste divisions, Bhils and Bhilalas have become more and more assimilated into the caste system, so that, in the plains, they are even described as a caste (Russell and Hiralal 1916: 293; Mayer 1960: 36). The economic subordination of Bhils and Bhilalas is mirrored in the hegemony of caste ideology; notions of hierarchy are more widely accepted in the plains. This is seen in the extent of 'sanskritization' in the plains — the claim to greater ritual purity by the adoption of upper caste practices and traits in order to move up within the caste system. One noticeable sign of such change among the Bhils and Bhilalas of Nimar is the rejection of the *kushta* (loincloth) in favour of the dhoti, the mode of dress of the upper castes.

Adivasis of the Hills

There is, however, a vast difference between the Bhils and Bhilalas of the plains and those of the hills. I shall argue that the geographical isolation of Bhils and Bhilalas in the 'fastnesses of the Satpuras' and Vindhyas has played a major role in enabling them to hold their own *vis-à-vis* the modern state and the caste system. This isolation was the consequence of a history of unequal struggle against the state and the market, experienced as various foreign rulers, traders and moneylenders. However, the particularity of hill ecology has, in turn, been central to the creation and maintenance of a community and an identity that is *adivasi*, separate and distinct from the Hindu system. Of course, identity is not static; it is continually made and unmade in interactions with different others. This dialectic is not a politically neutral one; it lies at the very heart of adivasi politics — their relationship with non-adivasis and other adivasis, with the state, with nature, and within the Sangath and Andolan. For the people we are talking about, identity *is* an issue.

The Historical Basis of Adivasi Identity

Bhil and Bhilala life in the hills shows that, processes of incorporation notwithstanding, they have a distinct identity today. This distinct identity is not a residual thing, a vestige of past glories, but is in part a creative response to processes of incorporation. In the discussion of the history of the region, I described how Bhils and Bhilalas who were ousted from the plains sought refuge in the forested hills. Even here they could not escape the exactions of the Rajput or British states; they had to pay revenue to the king. To do this, they had to borrow money from traders, to whom they were then obliged to sell their produce. The state had superordinate rights to the land they tilled and the forest they used. The best forests have now been felled for timber by the state, which has lost much of its interest in them, except as a way of feeding its customary corruption. The potential for nevad — taking over forest land for cultivation — has pulled the bottom out of the land market. The land is too unproductive for it to be worth the while of any self-respecting moneylender to take possession of it; a trader is better off profiting from the sale of adivasi produce. So the people of the hills are being left alone, more or less; their subsistence economy has little surplus that can be skimmed off.

Resource depletion, combined with their spatial isolation, has to a large extent allowed adivasis to maintain a structural distance from modern Hindu cultural institutions. It has given rise to a way of life which is unique to the Bhils and Bhilalas, Mankars and Naikdas — a life that sets them apart from the adivasis of the plains and from non-adivasis. Adivasis in the hills recognize that no one else can live like them; they take great pride in their toughness, the fortitude with which they can weather ill fortune, and the creativity by which they can bring forth sustenance from a seemingly inhospitable environment. This relation is conceptualized in myths of creation;[3] as a primordial link between God,

---

[3] See Appendix 2 for a transcription of the *gayana* (song of creation) of the Anjanvara Bhilalas.

the land and adivasis — a link that cannot be mediated by the temporal power of the state.

This understanding permeates the impassioned speech of adivasis threatened by the dam. In the words of Khajan, a Bhilala from Anjanvara:

God made the earth and the forest; then He made us, adivasis, to live upon the earth. Ever since we have come out of our mother's womb, we have lived here. Generation upon generation of our ancestors lived and died here. We are born of the earth and we bring forth grain from it. Governments live in cities and live on our grain. We live in the forest and we keep it alive. Governments and politicians come and go but we have never changed; we have been here from the beginning. The government cannot create the earth or the forest; then how can it take it away from us?

The Religion of Bhilala Adivasis

The link between nature and society is central to the religious belief of the adivasis of the hills. Of course, *all* peasant communities, living amidst the vagaries of weather, pestilence and prices, attempt to petition the powers of nature — rain and earth — with prayer. However, for Hindu peasants, animist beliefs coexist with a complex pantheon organized around the worship of Shiva, Shakti, or the incarnations of Vishnu. For Bhilalas, affecting nature's cycle is intrinsic to a cosmology that imbues *all* natural phenomena with spiritual life, so that the hills, trees, stones and crops actively intervene in people's daily life. The world is also populated by the spirits of ancestors — benevolently breathing over the shoulder of their progeny — and evil *daakans* (witches) who may possess married women. The village, which in the case of Anjanvara is congruent with the agnatic lineage and its wives, stands in a particular relation to a specific site and to its ancestral spirits who inhabit that site.

The conjunction of the natural, spiritual and social worlds can be seen in the collective performance of the most important Bhilala ritual — *indal pooja* (the worship of the union of the rain and earth which brings forth grain) — described at length in Chapter 7. The gayana, creation myth sung during indal, links the origin of the world to the river Narmada. The gayana, perhaps more than any other part of their religious complex, sets the

Bhilalas apart from the Hindus to whom the myth is entirely unfamiliar and who have no ritual context approximating indal.

Besides spectacular ceremonies like indal, there are many other beliefs that indicate the gulf between Bhilala and Hindu religion. Among mythologies, for instance, there is the completely different treatment accorded to the river Narmada itself. For Hindus, Narmada is, literally, 'the giver of bliss' (Guru 1983: 18). Born of the body of Lord Shiva, every stone of her bed is a miniature *shivalinga*, worthy of worship. The people along the banks hold her to be more sacred than the Ganga; local legends aver that Ganga herself must dip in the Narmada once a year. Ganga comes in the form of a coal-black cow and returns home quite white, cleansed of her sins. A Hindu proverb says that 'as wood is cut with a saw, so at the sight of the holy Narmada do a man's sins fall away' (Bhattacharyya 1977: 99). A Hindu ritual unique to the Narmada valley is the *parikrama*, the circumambulation of the river on foot. To perform the parikrama is to acquire great merit. All along the river, Hindus provide unstinting hospitality to pilgrims as they pass through their villages. Yet pilgrims complain that they are harassed, even robbed, when they go through adivasi villages along the river.[4] The ritual of circumambulation has no meaning for adivasis; pilgrims are fair game to them.

While Hindus on the banks of the Narmada have elaborate ritual calendars of worshipping the river, singing her praises, taking purifying baths in her water, asking boons of the goddess (to name only a few of many ways of showing reverence), adivasis seem remarkably casual in their attitude towards the river. The gayana sings of Narmada, but nowhere else in Bhilala culture is she deified in a way that matches her apotheosis by Hindus. Adivasis refer to the river as Narmada *mata* (mother), but whereas they have rituals for the propitiation of almost every natural phenomenon, little or no ritual surrounds the river.

Bhilala Peasant Economy

Reverence for nature mirrors the adivasi dependence on their immediate natural environment for material sustenance. Their

---

[4] Guru (1983: 40) reminds the prospective *parikrama* performer that the moral injunction to travel in austerity has a practical side to it in that it foils the marauding attempts of savage adivasis.

economy is completely based on the cultivation of land and livestock, and the collection of produce from the forest and river. While dependence on land and livestock would be a feature of, say, a peasant household in Nimar, the hill adivasis are distinguished by the extent of their self-sufficiency and the low level of their interaction with the market. Much that the people of Nimar must buy in the market is gleaned from the surrounding forest or taken from the river by the hill adivasis. The way in which production is organized in the hills is also quite different from agricultural practices in the plains. The productive community is defined, but not exhaustively, with respect to the village which consists of a set of relatively egalitarian relationships embedded in kinship. All households in the village have access to land and almost equal guarantee of subsistence. This point is discussed at length in Chapter 5.

The hill economy is marked by its communitarian character, which is in sharp contrast to the highly differentiated society of the plains. According to F.G. Bailey, 'the larger is the proportion of a given society which has direct access to land, the closer is that

society to the tribal end of the continuum. Conversely, the larger
is the proportion of people whose right to land is achieved through
a dependent relationship, the nearer the society comes to the caste
pole' (Bailey 1960: 13). No one is landless in Anjanvara; all but
two males belong to the agnatic lineage and inherit land. They
have also extended their cultivation to the nevad fields in the
forest. While ownership of these small landholdings is privatized,
they are worked jointly through institutionalized labour-sharing
practices such as *laah*.

Distinctions between Tribe and Caste

The emphasis on egalitarianism brings us to the differences between hill adivasis and the caste system. While earlier
anthropologists such as N.K. Bose, Verrier Elwin and von Fürer-
Haimendorf concentrated on the study of tribal societies, this
work came to be eclipsed during the fifties by a preoccupation
with caste as a system of organization. In keeping with the social
primacy of the caste system, academic attention has, since then,
tended to dwell on castes in India. As a consequence, 'tribes' have
languished on the theoretical peripheries such that it is difficult
to delineate adivasi social relations without comparing them to
those of the dominant caste system. In this section, then, instead
of continuing to use adivasi culture as the point of reference, I
shall invert the discussion and employ an ideal type of the caste
system as the touchstone.

The caste system is generally marked by certain distinguishing
features. Usually, caste boundaries tend to coincide with a system
of occupational specialization (Mayer 1960). Castes are stratified
on the basis of the ideology of purity and pollution, different
degrees of which are associated with different occupations
(Dumont 1970). Within the endogamous caste group, people try
to bring about hypergamous unions so that the marriage of a
daughter improves a family's status. The ideology of caste is
acknowledged by all groups, including low-ranking ones, who
challenge their place in the hierarchy even as they do not question
its overall legitimacy (Chatterjee 1989). I shall discuss each of
these features in turn, and examine the extent to which the hill
adivasis fit into this scheme.

The caste system in a village defines the traditional division of labour. It is an ascribed system of occupational specialization which extends beyond Hindus to include Muslims and other religious groups as well. For instance, not only are Brahmins priests and Vaniyas traders, but Balais (a Muslim caste) are weavers (Mayer 1960: 34). Interdependence in the village is structured along the lines of this division. This contrasts with the structure of a hill adivasi village such as Anjanvara where there is little occupational specialization beyond that by gender. Even 'specialists' such as the patel (headman) and pujara (priest) are farmers first. As the village consists of the lineage of one clan, there is little differentiation into ascribed occupations. All tasks (other than those demarcated by gender) can be performed by anyone. For instance, while there is one 'official' budva (shaman), several other men possess such skills and may be called upon to fill that role.[5] As G.S. Aurora observes, even when 'the economy of the adivasis was in transition from a subsistence to a market-oriented economy . . . [i]t was not developing towards the model of inter-caste village economy but was rather in the process of integration into the wider commercial economy of the region' (Aurora 1972: 18).

Castes are arranged into a hierarchy based on the purity and pollution associated with different occupations. This hierarchy is expressed through strict proscriptions about eating and marriage. Bhilalas in Madhya Pradesh also rank themselves according to this scheme and consider Mankars, Bhils and Naikdas their social inferiors with whom they do not eat or marry.[6] Bhilalas in Aliraj-pur have vernacular terms for pollution (*osotne*) and maintaining caste taboos (*chhint mekne*). It is curious, though, that ranking behaviour appears to be absent among the Bhilalas on the Maharashtra side of the Narmada, even though there is intermarriage between the Bhilalas on either side of the river.

Where it does exist, ranking seems to owe a great deal to Hindu

[5] In larger villages, there may be greater specialization. For instance, the Naikdas of Kakrana rely more on fishing than the Bhilalas, but the contrast is not that of absolute difference, but of degree.

[6] Besides such ranking, there are few differences that set the Bhils apart from the Bhilalas, yet either community is careful to distinguish itself from the other. Where Bhils and Bhilalas reside in the same village, they tend to live in separate *phalyas* (hamlets). They speak different dialects. Generally, Bhils are poorer than Bhilalas and depend more on collecting forest produce to earn a living.

influence. The Bhilalas, who are said to originate from the miscegenation of a Hindu caste (Rajput) and Bhils, span that cultural
divide. Of course, upper-caste Hindus generally lump adivasis
with untouchables and will not accept food from their hands.[7]
The adivasis are doubly misused — Hindu officials who raid their
village will carry off their hens and force them to give flour for
rotis, but insist that it be cooked between leaves (*pannia*), instead
of in adivasi utensils, so that their polluting touch can be avoided.

The acceptance by Bhilalas of a caste hierarchy has some ambiguity. In the Hindu case, the system of stratification is sanctioned
through myth so that Brahmins came forth from Lord Brahma's
mouth, Kshatriyas from his arms, Vaishyas from his thighs and
Shudras from his feet. Such a gradation does not exist in the
Bhilala myths of creation, which describe Bhilalas, Bhils, Dheds,
Chamars and Mankars as all coming out of the bodily wastes of
the original couple Dhedya and Dumbda. The myth enjoins
endogamy within each of these groups but does not structure them
into a hierarchy. For Hindus, the notion of caste as a station in
life is legitimized by the philosophy of karma which, through the
transmigration of the soul and the cycle of rebirth, holds out the
hope of a higher status in the next life. No such justification for
the present is found among the Bhilalas.

Adivasis say that, with modernization, just as Hindu caste
taboos are changing, ranking by Bhilalas is gradually losing its
rigidity too. When people travel to the haat, they eat and drink
in shops where they cannot be sure of the caste identity of their
neighbour. The Sangath has also been instrumental in creating a
common adivasi identity by mobilizing around the slogan *Amu
akha ek se* (We are all one).

In the caste system, women guarantee the ritual purity of the
household. However, as long as a Bhilala woman marries a Bhilala
man from a clan other than her own, she is allowed great flexibility
in the choice of spouse. In this regard, Bhilala marriage customs
are quite different from those of Hindus. In the villages of the
Nimar plains, especially among the dominant Patidar caste, marriage is a privileged means of improving one's status, and is

[7] They *will* accept water, though. In Nimar, Patidar households have a finely
calibrated system of spatially-demarcated untouchability. While an adivasi can
enter the open courtyard of the house but must not step onto its raised, covered
portion, a Harijan will not even be allowed past the door.

articulated with the caste system (Pocock 1972). In this hypergamous model, the guiding ideology is that of *kanyadaan*, the 'gift
of a virgin'. A family acquires a higher ritual status by arranging
the marriage of its daughters, each endowed with as large a dowry
as possible, with a higher family. 'The gain of status alone com-
pensates the economic losses' of such a union (Deliege 1985: 118).

Deliege remarks that

marriage among the Bhils is by nature contrary to Hindu marriage . . . [8]
The gap between the hill and plains sections is, here again, wide. By
aligning themselves with the Hindu castes, the plains Bhils completely
modify [the] nature of their marriage and fit the traditional pattern [of
the Hindus]. In the hills, isogamous unions are not second best, under-
valued solutions but command the whole institution of marriage. The
question of status is absent from the matrimonial transactions; a 'good
marriage' does not carry much prestige. Brideprice among the Bhils is
not a shameful practice but a rule and hence it is opposed to the Hindu
ideology: unlike low Patidars or Rajputs, Bhils are not embarrassed by
paying brideprice and this difference is indeed fundamental (Deliege
1985: 119).

Thus, among the hill adivasis marriage is marked by considerations of sentiments other than those of status. There is much more
autonomy in the choice of spouse; elopement and abduction are
socially sanctioned. The festival of Bhagoria in February–March,
when adivasis gather at the haats to select their partners celebrates
that freedom of choice. Marriages are sealed by the payment of
brideprice to the woman's father; they are broken when she leaves
her husband, who then tries to get his money back. The ease of
divorce also sets the hill adivasis apart from the people of the
plains. Finally, the notion of 'the gift of a virgin' appears ludicrous
in a society that does not place a high value on the chastity of
unmarried women. Sexual liaisons are easily acknowledged and
people live together before wedding rites are performed. All of
which is quite contrary to the Hindu norm.

The extent to which a community accepts the ideology of
hierarchy indicates the degree to which it accepts incorporation
into the structure of the dominant Hindu caste system. By ranking
themselves *vis-à-vis* other communities, hill adivasis do conform
to this caste characteristic. Yet in many significant ways pertaining

---

[8] Deliege treats the Bhilalas as a subsection of the Bhils.

to religion, economy and social structure, their beliefs and practices are in sharp contrast to the ideology of the caste system. Deliege remarks that 'those practices which are considered as inferior in the low castes are in fact the rule, the admitted and encouraged pattern, among tribal people . . . [W]hen a tribal section becomes more Hinduized, it does not necessarily change its habits and customs at once, but it becomes ashamed of them and tries to deny or conceal them. This sense of inferiority . . . is perhaps the best indicator of a transformation of a tribal group into a caste since it denotes a loss of its dignity, its pride, its sense of equality and its independence' (Deliege 1985: 160).

The Negotiation of Hinduization: The Mata Phenomenon

The ongoing conflict between adivasi beliefs and practices and those of the dominant caste society can be seen in the periodic passage of the *mata* (mother) through the Bhil region. In Alirajpur, they say that, in 1992, the mata came from Pavagarh in Gujarat, then swung towards Meghnagar in north Jhabua, came south to Alirajpur, and from there travelled east into Dhar. In four months she passed like a wave from one village to another, possessing people and inducing others to follow her commands.[9]

Everybody knew when the mata entered the village of Anjanvara. The people from the neighbouring village of Sirkhadi came to the boundary between the two villages and left a *sivdu*[10] — the assorted paraphernalia that marked their devotions to the goddess. Each household in Sirkhadi had given an earthen pot, a coconut, incense sticks, vermilion powder, a comb, a small mirror and ribbons for the mata. The pots were not the black clay ones particular to Alirajpur; they were the brick coloured pots that are widely used in north India. When the sivdu was deposited on the border of Anjanvara, the goddess was said to have entered Anjanvara.

Through word of mouth, and by noting what had been done in other villages, the people of Anjanvara knew what to do. First there was the

[9] Why did the mata appear in our midst? People said a great sin was committed; a man had incestuous relations with his mother and his sister. Another explanation attributes the coming of the mata to another sin — that of unnatural sexual intercourse between men and animals. A buffalo gave birth to a boy and a cow gave birth to a girl.

[10] '*Sivdu*' is the term used in Anjanvara for what is elsewhere called a '*peeda*' (see Chapter 7).

palni (set of observances). No food was cooked during the day; only tea was drunk. No meat, fish or eggs were eaten at all. People stopped drinking liquor. During the four days of the palni, Bhilala rituals that involved animal sacrifice and the ceremonial drinking of liquor were suspended. Women and children did not use their customary scented snuff or the roasted tobacco that they rubbed into their teeth, but started smoking bidis like the men. The start of the palni was marked by all the men going off to bathe in the river. The bathing was repeated frequently. People dressed differently for the *mata*, the men wore only white — white Gandhi caps, white shirts and white shorts of thin cotton. Women started wearing white *sarees*, tied in the Gujarati style.

Then the possession started. Slowly, those people in the village who were known to be spiritually sensitive and who had a propensity for possession, started shaking, rolling their heads so that their hair flew about, convulsively writhing on the ground while moaning and mumbling strangely. They felt the mata as an unbearable heat inside their bodies and, in their trance, kept demanding water to cool them off. Other people would hurry to them, bearing water in brass *lotas* (small metal water pot) which they would sprinkle on their heads and face. While possession is usually limited to Bhilala men, the mata entered women too, endowing them with the power of clairvoyance, enabling them to identify the evil spirits or *daakans* in their midst. The mata spoke through the possessed, forecasting the future, directing people and events for her proper propitiation.

Then a sivdu was taken out through the village, just as it had been done in Sirkhadi, with each household offering what the mata asked for — an earthen pot, a coconut, incense sticks, vermilion powder, a comb, a small mirror and ribbons. All these offerings were together left on the border with Sakarja, and the mata was said to have passed into the next village. The people of Anjanvara collected money and a goat, which they sent to the goddess at Pavagarh. Then, in a procession, they left the village to go to a small Hindu temple on the bank of the Narmada. Dressed in white, their foreheads smeared with vermilion, with coconuts, incense sticks and a little rice in their bags, carrying lotas with water to cool those possessed by the mata, they walked and danced their way to the temple.

About two hundred people are gathered near the river, and drums and pipes are playing. All in white, the crowd pulses with the fervour of the mata. Women are dancing with swords held aloft, their hair wildly loose upon their shoulders and over their face. It is like a chain reaction — one person starts shaking violently and sets off the rest so that, in a while, the possessed are all dancing crazily, moaning that the heat be taken off, while others are rushing to fetch water and sprinkle it on those possessed. Possession ebbs and flows. For a few minutes there is an air of normalcy as people walk towards the temple, then the mata surges forth once again and there is renewed frenzy. In this way, with frequent stops and starts, they go to the temple to take off the *bhaar* (spirit) of the mata, where they break coconuts, light incense sticks, smear vermilion on the idol and offer it rice grains. Then they bathe in the river and are said to have cooled off completely. On the way back, they offer coconut bits, rice grains and vermilion to the various village deities, and are done with the mata.

But the mata came back about four times over the space of four months, meandering through the villages of Alirajpur. In each successive wave, the rites of propitiation were modified slightly. Each time, the villagers observed the palni faithfully, sending off whatever the mata called for, fasting during the day, going to the temple and generally being *sukla* (pure, clean, in the Hindu sense of the antithesis of polluted).

In many ways, the coming of the mata clearly appears to be a movement towards greater Hinduization. The mata's palni calls for the cessation of adivasi rites of worship. The customary sacrifice of animals to propitiate adivasi gods is no longer allowed. No liquor can be used to worship the earth or the ancestors. Even traditional adivasi marriage practices — brideprice, abduction and elopement are proscribed. In their stead, adivasis have to adopt sukla rites, imitating the practices of upper-caste Hindus. Fasting, the emphasis on water and purificatory baths, the use of coconuts, vermilion, incense and rice for worship, going to a temple, are all alien to adivasi culture. The prohibition on eating meat and drinking liquor is a wrenching sacrifice for a community that holds both these things dear. Under the circumstances, the passage of the mata seems to be a move towards Hinduization.

Yet the mata has been made to serve all kinds of ends. I heard tales about the mata, speaking through girls who were possessed, commanding their parents to approve their choice of spouse. There were accounts of women moved by the mata beating up corrupt government officials and travelling in buses without paying the fare, and so on. Khemla, the president of the Sangath was also possessed by the mata and prophesied that a time would come when all forest land would belong to the adivasis (see Chapter 9). But it was generally held that a command or prophecy that profited the possessed person was suspect; the mata would only speak *through* the medium, never for the personal gain of the medium.

In a significant departure from the Bhilala practice, the ability to be possessed by the mata was not limited to men. Women, who are usually regarded as morally weak, susceptible only to possession by evil spirits or daakans, were during the mata days visited by the goddess and granted clairvoyant powers. Whereas women are completely excluded from Bhilala rituals, even to the extent of being barred from cooking for the communal feasts, the coming of the mata gave them a privileged place as interpreters of the divine.

David Hardiman (1987) describes a startlingly similar movement that occurred in south Gujarat during the 1920s among adivasis. In that case too, the coming of the *devi* led to the observance of new rules of purity such as temperance, vegetarianism and cleanliness. However, in south Gujarat, the devi commanded people to boycott Parsi landowners, traders and moneylenders who, through their control of the liquor trade, had bound many adivasis into indebtedness and had appropriated their land. At the same time, the new rules of purity were also linked to the call of Mahatma Gandhi and the nationalist movement. According to Hardiman, the devi movement in south Gujarat was part of an attempt by adivasis to assert themselves against politically dominant classes.

Could this argument be true of the mata wave that swept recently through Alirajpur?

Hardiman asserts that the devi movement was not an attempt at 'sanskritization' (1987: 157–60) since the adivasis did not claim a higher *caste* status for themselves. This is also true of Alirajpur where, in adopting upper-caste practices, adivasis did not aspire to a higher status within the caste system. While they did not see themselves as becoming Hindus, they *were* moved by the notion of *sudhaar* (improvement) — the betterment of their moral selves. In this way, the mata wave could be thought of as the 'democratization' of upper-caste values. However, in the Alirajpur case, it must be emphasized that the values endorsed by the mata were a departure from the adivasi norm and involved the adoption of upper-caste Hindu practices. To the extent that adivasis gave up drinking liquor, eating meat and worshipping their gods through animal sacrifice, they *were* repudiating their adivasi identity and embarking on a round of Hinduization.

The mata also undermined adivasi economy in two ways. First, it led to long periods of enforced 'idleness' when people neglected their agricultural labour for almost four months in order to follow the palni. Second, the mata greatly profited non-adivasi traders who sold the clothes and items of worship that the mata called for. The bazaarias generally looked with benign approval upon the religious fervour sweeping the adivasis for they regarded this as sudhaar, the emergence of the 'backward' adivasi from the slough of savagery.

To see the mata as a move towards 'democratization' is problematic for another reason too. While the mata endowed women as well as men with the powers of prophecy, giving them a spiritual status that was previously denied to them, very often this power was turned *against* other women. Those possessed frequently identified other women as daakans who were then put to death. Usually, older women — widows, those who are childless — tend to be identified as daakans. Since these women are considered to be 'marginal', different from the accepted norm of the married woman blessed with children, their existence tended to be perceived as a threat to social wellbeing.[11] The spurt in the killing of women after branding them as daakans shows that if the mata brought along a 'democratizing' tendency, it did not encompass all women. In fact, it *worsened* the social position of those women who were already badly off.

While the mata has been a movement towards Hinduization that involved imitating the dominant castes, in many ways its success can be attributed to the way in which it resonated with existing adivasi culture. Significantly, those adivasis who have become *bhagats* (people subscribing to upper-caste practices and completely rejecting adivasi rituals) contemptuously dismissed the mata as mass hysteria to which their 'backward' brethren were deplorably prone. The rise in the killing of daakans that occurred during the mata wave combined new religious justification with an old adivasi practice, namely the identification of daakans. Certain aspects of worship, such as taking out the sivdu, were familiar to adivasis who routinely employ this as a method for expelling evil spirits from the village. There is also a continuity

[11] Since the mata is worshipped with vermilion, feminine adornments and other auspicious offerings, she is clearly thought to be a married goddess, the opposite of the evil spirit of the daakan who possesses widows and the childless.

between the appearance of the mata through the person of a possessed medium and the common adivasi practice of possession during religious ceremonies (see Chapter 7). Thus the mata builds upon adivasi rituals and incorporates them into a Hinduizing movement.

How can we be certain, though, that the reverse is not the case, and the mata is not the incorporation of Hindu elements into adivasi religion? It is significant that, while adivasis seem to be obeying the commands of the mata of their own volition and 'external' pushers of these beliefs are apparently absent, adivasis are conscious that following the precepts of the mata involves giving up their own cultural traditions. While they acknowledge the desirability of sudhaar — of improving their moral selves by adopting the sukla practices of upper-caste Hindus — they are anxious about incurring the wrath of the adivasi gods. At the same time, almost everyone felt economically encumbered by the claims that the mata made upon their time and their money. These conflicting tugs of different religious traditions were also experienced in Anjanvara.

All the men gathered in a village council to decide what to do about the palni which will reach Anjanvara soon. They had faithfully fulfilled the demands of the first palni. Now they have to determine the logistics of the next one. This time they say that they will make do with white caps and coconuts, no more. This is still a considerable expense for them. Caps cost six rupees apiece. Someone had bought one a few days ago from Maharashtra across the river; it is passed around and examined. Coconuts cost six rupees apiece too. Each household will need at least three coconuts. In all, the village needs 179 coconuts and 45 white caps for the men and older boys. Trying to cut corners, people say that they will make do with the white shirts and lotas that they already own. There is a brief discussion on whether a white shirt with thin blue stripes will be acceptable for the palni or not. They also calculate how many people in the village can be expected to get possessed and what they will require. Two people are deputed to collect money from each household and go to the haat to buy the caps and the coconuts.

In determining how to get by with minimal expenditure without offending the mata, people are moved by two sets of considerations. On the one hand, an inadequate palni will rouse the ire of the goddess; perhaps people will be visited by pestilence and disaster. On the other hand, how can one give up the old gods and the old ways forever? How will crops ripen without the offerings of goat and chicken? . . . Without animal sacrifice and offerings of liquor the adivasi gods of nature will not smile upon the people and their crops. So, for the moment, adivasis somewhat reluctantly do what the mata demands, but they know that they will revert to their old ways once the period of the palni is over.

Most people regarded the mata as a goddess who had to be propitiated so that life could return to normal. Any attempt to interpret the mata wave as Hinduization must be qualified ultimately by trying to assess the *permanence* of the changes in beliefs and practices wrought by the goddess. If the mata only passes through the region, bringing about a short-lived change in people's religious lives, then it is difficult to say what its lasting impact will be. If people, conscious of the implications of this change, weigh all the consequences and choose to adhere to their traditional religion, then we see the strength and resilience of adivasi culture.

Distinctions between Hill Adivasis and Bazaarias

The refusal, on the whole, of the hill adivasis to repudiate their indigenous beliefs contrasted with the surrender of their counterparts in the plains. For the people of the hills, another and probably more salient system of social differentiation occupies the mind. This is the distinction between hill adivasis and bazaarias (people of the town) — non-tribal traders, moneylenders and government functionaries. However, the economic and political domination of the non-adivasi population makes the haat and the tehsil headquarters a cultural frontier where the hegemony of modernization and caste ideology is sought to be extended to the hill adivasis. This has resulted in an ongoing dialectic. On the one hand, the subordination of the hill adivasis to bazaarias who control the market economy and the state has made the values of the dominant culture appear desirable and to be emulated, giving rise in some cases to bhagat movements towards Hinduization. On the other hand, the juxtaposition of adivasi and non-adivasi cultures has thrown into relief their differences, which have been seized upon by those aiming to unite adivasis for political ends. The bazaaria view of the adivasis is illustrated in a statement made by the Sub-divisional Magistrate of Alirajpur, an IAS officer,[12] as recently as in 1993:

Alirajpur is a remote sub-division of Jhabua district having more than 85 per cent of population coming under Schedule Tribes. People of this area usually carry bows and arrows and falies [sickles] with them and they are always under the influence of country liquor or toddy. On very petty issues they quarrel among themselves and attack each other with sickles, arrows and stones fired from Gofun. These stone projectiles are very deadly and a person may even die if hit by them. Most of the deaths are caused by arrow, Gofun and sickles in this area.

We see that, in some ways, very little has changed over the last hundred years. The modern administration echoes the words of a colonial officer, the Superintendent of Alirajpur, who referred to 'the ignorant and uncivilized Bhils of this state . . . [whose condition] is too well known to be described here. Because on simplest provocation and trifling causes they kill each other' (NAI 1896).[13]

According to bazaarias, adivasis are *gande*,[14] who will never improve. Aurora narrates that, 'One of the common remarks [in Alirajpur] is "the tribal is a primitive Englishman" . . . A schoolmaster said: "Like an Englishman the tribal lives in houses separated from each other. He does not like to be interfered with by the neighbours, preferring to live with his wife and children and working on his own piece of land. He does not use water to cleanse himself after defecating but instead uses a stone. Again, like an Englishman, he always keeps his wife with himself. His morals are also similar to the Englishman's. Young men and girls can have sexual relations without too many restrictions, and like the English people they have love marriages more than arranged marriages" ' (Aurora 1972: 159). Implicit in this analysis is the contrast with the bazaaria's own life and his contempt for the life of the adivasi.

This contempt can be echoed by adivasis in a self-deprecating way that mocks and confounds the best intentions of the state.

[12] This statement was made in a deposition to the Supreme Court (see Epilogue).

[13] *Rules for Extradition of Criminals Between Chota-Udaipur and Alirajpur*, 7 July 1896. From the Superintendent, Alirajpur, to the Political Agent, Bhopawar. Bhopawar Political Agency. S. No. 5. 1/67/1896. NAI.

[14] It is hard to capture the sense in which *gande* is used; it falls somewhere between delinquent, loutish, maliciously mischievous.

Aurora has a nice anecdote about a medical team's visit to a village which shows how the ironical appropriation of bazaaria attitudes can be a subtle form of passive resistance:

The malaria team came to Bamanta and wanted to take blood samples of the Bhilalas. The children were brought together by the school master. They came not knowing what was up. The malaria worker tried to explain to some of the men who had also gathered. 'This needle will cure you of fever' he said. Soon as the children heard that they were to be pricked all of them ran away. The worker was dejected. Meanwhile, Jwan Singh brought toddy from atop a tree. 'They are all *gandes* [*sic*]', one of the [team members] said. '*Ho*' [yes] replied Jwan Singh. 'Government is spending so much money for your benefit, it is all a waste'. 'We are gande', replied Jwan Singh. 'Hey! Why don't you wear something over your buttocks, don't you feel shy'. 'We gande people don't look well in a *vanya's* dhoti; they are too delicate for our use', said Jwan Singh' (Aurora 1972: 208–9).

Despite their respect for the bazaaria's power, adivasis by their dress, speech and lifestyle show that they do not fully accept the moral hegemony of the bazaaria. On the contrary, they believe themselves to be superior in many ways and strive to maintain that distinction. This is exhibited in the differences in the organization of their lives and works, and the values upon which that is based — boundaries that they continue to maintain despite overwhelming pressures to the contrary. Bhilala identity is not taken for granted, but is actively constructed and passionately defended. As the next chapter suggests, adivasi identity is based on their sense of community, a concept that endures over time and space. Approaching an understanding of the meanings of community is essential for an appreciation of adivasi consciousness.

5

Community and the
Politics of Honour

In the previous chapters we saw how tribal history has been a long struggle against the state's appropriation of political power, a process marked by the progressive alienation of adivasis from nature. The domination of the state and the market has been experienced through the presence of bazaarias who, besides being government officials or traders, have also acted as bearers of Hindu caste ideology. The adivasis of the hills have, to a large extent, escaped the fate of incorporation into the caste system which befell their people in the plains. Here I shall describe the adivasi community in terms of the set of relations which define it and give it corporate form. These relations contain contradictions from which emanate many of the tensions of daily life, affecting the organization of work within the household, the village and the larger world. These relations also shape the politics of the community.

A Day in Binda's Life: 11 April 1991

Before it is yet dawn, before the first cock has scarcely crowed, my sleep is broken by the steady whine of the grindstone turning. In the dark, Binda has risen, splashed some water on her face, and set to work grinding flour. I know it is hard work; each time I have tried I have given up after a few minutes, unused as I am to the steady rhythmic exertion it calls for. Binda does it for two hours. She has to feed nine people so there is a lot of *juvar* to grind. The grindstone is an inefficient device too, doing tiny amounts at a time. Things are a little easier for Binda now that it is summer and she can cook *raabdi* — a thick soup of boiled, fermented grain — once a day and *rutu* — flat unleavened bread — once. Otherwise, flour for two sessions of rutu means twice as much work.

Binda has four children — her eldest daughter Dokli is about ten; she is old enough to care for the baby, sweep the house, help her mother cook, fetch water and collect wood. Next is Baki, who doesn't really like having to go every day with the cattle or the goat; she wants to stay at home and play, making mud pots or feeding the baby parakeet. So Binda is often angry with her and yells and nags. Then Baki looks sulky and goes off to the river, away from the house. Radya, the elder son, is about six years old. When he plays outside, he darts into the house every now and then yelling, 'Ayaa! Khanyaa!' (Mother! Food!) When I tease him and echo 'Ayaa! Khanyaa!' after him, he grins and hangs his head sheepishly. He is cheeky and a little spoilt; he cries if he can't get rutu at every meal and his parents tend to indulge him. Binda not so much, but her husband Khajan takes time out to make a bow for his son and to play with him. But the person on whom they shower the most affection — hugs, kisses and baby talk — is the youngest, one year old Lotya.

Lotya is much more happy now and active. About two months ago, his entire head was covered with infected boils and sores that he'd scratch and worsen. He also had what looked to me like chronic conjunctivitis — eyes gummy and swollen every morning. So he'd cry and cry during the day and much of the night. And Binda would have to be up, rocking his cradle for hours together. Now that Lotya is free from his ailments, Binda can stop worrying and enjoy his infancy, his smile as he responds to her, his mother. 'Four of our children have died, two after they were old enough to walk', she tells me. So the present situation of four healthy children is about as good a life as she hopes for. She doesn't want to have more children, but doesn't want herself or her husband to go in for the operation. People she knows who have had it have lost their strength and can no longer work as they used to. She knows of no other method of not having children; she feels that her fertility is a part of her fate.

After the grinding of the flour is done, Binda goes to fetch the first pot of water of the day from the river, just as the sun is coming up. Down to the river, then climb up again, quickly because there is so much left to do. She gives *helo* — cold food from last night — to the children who squabble for a place by the kitchen hearth, and tears off a piece of rutu for herself. Then the cattle and their one goat are let loose to join other families' herds for grazing. Binda shouts, 'Baki! Other people's goats are past the *patel's* house. Hurry up!' 'Why can't Radya go today?' asks Baki. 'He's too small. Go on! Take the small sickle and go.' Armed with the sickle for lopping off leaves and branches, Baki goes off grumbling. Lotya wakes up and starts crying. While feeding him, Binda lights a fire to make some tea and it is quickly drunk by all.

Now that the work in the fields is done, Binda doesn't have to cook

early in the morning. Otherwise she'd have to quickly make *ulhan* (a dish of cooked pulses, vegetables or meat, usually wet, with which one eats rutu) and rutu for the men to take up to the fields in the hills where they farm on encroached upon forest land. Now she can use the morning to collect firewood from the trees on the hills. The summer is time to build up your store, because when the rains come and it is time to plant, there will be too much work (besides, the wood will be damp) to leave time for fetching wood. She comes back with a headload of dry wood — mainly teak and some *halai*. Now she also burns *tuvari* stalks left over from the winter crop.

Binda fetches water again and starts cooking the midday meal which is usually ulhan — either dal or *maasan khato* (fish cooked in *kaanji* — a fermented liquid of flour and water) — and rutu. Today the ulhan is tuvari cooked with bhanjan leaves that she has plucked from a vine near the river. Men trickle in from work and, with the help of Dokli who cleans the brass plates and serves water, Binda gives them food. They have only three plates, so they wait for their turn to eat. As they squat on the floor, Khajan asks his brother Dhemchya, 'Did you take the bull over to Budhya's?' Dhemchya grunts assent between mouthfuls of food. Binda urges me to eat more, 'You eat so little. I couldn't even shit once a day if I ate as little as you eat'. 'But you work so hard', I point out, 'I just sit around. I don't *need* to eat as much as you eat'. We have this conversation every day.

Binda finally eats and feeds Lotya too. Free for a while, she roasts some tobacco on a piece of smouldering *saan* (dried cowdung), crumbles it and rubs it into her teeth. She only does this about thrice a day; Khiyali, her neighbour, who also eats tobacco in this way but more often, has the join between her teeth and gums stained black. When she laughs, I see white teeth outlined with black and then pink gums — a strange effect. Khiyali has brought her baby Mirli with her. Mirli can crawl and she moves towards Lotya who sits sucking his mother's necklace. I notice that Mirli has sores on her hands and feet that she keeps scratching, probably some of the scabies that's going around. I launch into my health worker role again, 'Khiyali, why don't you make that paste of neemda and turmeric like I told you to do? And keep Mirli away from Lotya or he will get it too'. Khiyali sighs and goes off to borrow turmeric. I am conscious that I am nagging, but no one listens to me otherwise. And there is a full-fledged scabies epidemic raging across the river which puts Anjanvara at risk.

Binda fills a vessel with water and carries Lotya off to that part of the house where the cattle are tied and where Lotya's umbilical cord is buried. Making Lotya lie down on her outstretched legs, she bathes him with a bit of soap. He doesn't like it and yells his protests. Then, murmuring fond loving things to him, Binda dries him off, feeds him and lays him in the cradle to sleep.

It is really hot now, the sun raining down its merciless heat. Binda covers her head yet again and walks down to the river, moving quickly to lessen the sting of the burning sand on her feet. As she walks to the water, her saree is a speck of brilliant blue against the shimmering wide valley, the shadeless rocks through which the river flows. Another pot of water filled and carried back up the hill. Yesterday Binda had split some bamboo into thin strips and soaked it in water to make it pliable. Today she takes some anjan fibre and twists bits of it on her thigh to make it into rope. Once that is done, she fetches the bamboo strips and bends them around the rope made taut by being tied to a post, twisting the strips in half and securing them with the rope. Then, once all the strips have been attached in this way, and are lying in a row, she rolls them up, ties them with the loose leftover end of the rope, and she has made a *kharatu*— a broom for sweeping the bare area outside the house.

Then the children are back from the river. Radya bounds in, 'Ayaa! Khanyaa!' Binda hands him a piece of rutu but he pushes it aside, 'Give me egg with it'. Khajan comes in from the nearby *maandava* (an open structure of wood poles, the roof of which is used to store fodder) where he has been smoothening beams for the house they will soon start building. He wants food as do his two brothers who live with him. Binda makes more rutu, quietens Radya with a bit of fish saved from the last meal. Lotya is up and crying and Khajan scolds Binda, 'Oh, it's a sin against god to leave this poor helpless baby thirsty. It is so hot that even a man drinks water every hour. How can a little baby get by?' Binda argues back, 'You were outside. You didn't see, but I just *gave* him a drink of water' Binda and Khajan are happy together. They talk about their day, about events and people that concern them both; they joke a lot and share their joy in Lotya. What they don't share — and this is an area that Binda accepts even if she finds it inexplicable and closed to her — is Khajan's political work as an activist. He is away from the village a lot, attending meetings; he has been to places as distant as Delhi and Bombay. Binda, on the other hand, has only travelled to villages where she has relatives and to some nearby haats.

'When Dokli's father (this is how Binda refers to her husband, not saying his name) goes to a meeting, I have no way of knowing when he will come back. During the Narmada Sangharsh Yatra he was gone for almost a month. When he is gone, all the farming work which must be done and which can't wait, has to be done somehow. There was the time when grain had to be brought down from the hills to be stored and there was no one to take care of it. His brothers called a *laah* on his behalf so men from different houses came over to lend a hand with the work and I killed a chicken for them afterwards. They did get all the grain home in the end, but it was really delayed and I was so worried that I'd lie awake at night thinking about it. Because of Dokli's father's roaming, ours was one of the last *udvaas* (round, high stacks of sheaves of grain which stand near the house) to come down for the grain to be threshed'. Binda spent most of that week in the *kholo* — the round clearing where the bullocks are made to walk on the ears of grain — with a *hupda* (a large fan-shaped platter of woven cane) poised over her head, waiting for the surges of wind which would separate the chaff from the bajra.

Now that winnowing is done and the grain stored in the bulging baskets of bamboo which they call *mutis*, Khajan is busy collecting beams for his new, enlarged house. His brother Shankria, who lives with him, is weaving a large screen of hiyali that will be one wall of the house. Binda is anxious that there is so much to be done for the house and they still don't have all the wood they need. If Khajan didn't travel so much, they would be much further along by now. Not that Binda resents Khajan's political activity; as much as she understands or is interested, she feels that he is doing something necessary; she just doesn't like the worrying situations it sometimes puts her in.

Binda also fights. But not in the way Khajan does, going to meetings, demonstrations, uniting with people from other villages. 'Last year, in the summer, the surveyors came to Anjanvara to see what would be submerged by the dam. They had forty policemen with them, with guns and on horseback. First they went from house to house and made us all go to the patel's house. When we asked why, they said they were going to hold a meeting. But then they shut the door and put a guard outside. Then some of us managed to slip out from the back. When we saw what they were doing, we ran forward and snatched away their measuring tapes.' For this they were mercilessly beaten by the police. Binda too was hit with lathis on her head and her back. Her niece Sevanti fell down and policemen trampled over her body. Today someone has come back from a meeting at Attha with the news that some more villages would come under water with the rains this year because of continued work on the dam. What would they do when the waters come? Move up to the hills where their fields are, seeing how their present village would be drowned deep in the reservoir. So the spectre of submergence, of having to rebuild their lives under even harsher circumstances, looms large in the distance.

Binda probably doesn't want to think about it, but if the dam is built and they do move up to their fields in the hills, an ocean will separate her from her natal village Arda. At present, Arda is across the river. Binda can't swim very well, but Khajan makes her a rough raft on which

she and the children float across to the other side. Binda's parents and
her five brothers' families live in Arda. When she goes there, she can
get a short respite from her responsibilities, get news about all her
relatives and spend time with her parents. The children look forward
with great excitement to visits to Arda. Their grandparents fuss over
them; they get to eat the choice things of the season — groundnuts,
fruit of the *temru*, mangoes. And there is usually a special meal where
a chicken is killed.

For Khajan too, Arda has been a source of support. When they didn't
have much land and when the rains failed, he farmed in Binda's village
with the help of her brothers. He has also cut wood from their forests
and carried it on his head to Dhadgam to sell. Even now, when he and
his family are much better off, their situation more secure with stocks
of grain to last them through the year, the help from Arda is valuable.[1]
Khajan has got a lot of wood for his new house from there. And Binda's
parents always send their daughter back home with a gift — a bagful
of *muhda* flowers, a bamboo basket woven by her father, or some roasted
gram. If the dam is built, this network of affection, of caring and help,
will come to an end and Binda will be marooned on the mountaintop.
If the people of this village were given land in Gujarat (as the government
claims it will do) Arda would be distant, visiting there a dream. But
Arda is in the submergence zone too. What will happen to Binda's relatives, will they all be scattered? Will Arda, as Binda has known it in her
girlhood and beyond, cease to exist just as Anjanvara will? Will Binda
lose the two places that define her life, her sense of location, her identity?

It is afternoon now and thirsty children wandering in and out have
finished almost all the water in the pots. So Binda goes down to the
river again, yelling 'Dokli! Mind Lotya! Sing him to sleep'. Baki is again
reluctant to go with the goat for its second forage of the day so Binda
has to shout to make her drag herself out of the house. Khiyali from
the house next door is also going to fetch water, so Binda picks up
Lotya's soiled clothes to wash after she bathes. Down by the river, Radya
is in the water with other boys his age, busy fishing and giving each
other shrill instructions. Some have caught a few fish, all no longer than
a man's finger and hardly enough for a meal. Radya hasn't caught any.
Today Binda doesn't have time, otherwise she also goes to catch fish
with Khiyali, using her *saree* as a net. She catches minnows, lots of tiny
silvery slivers.

[1] Even though the patel of Anjanvara — and therefore the entire village in
a way — has a feud with the people of Arda and has broken off ties with them,
Khajan has risked the displeasure of his clansmen and has paid a fine to individually settle his part in the dispute and maintain cordial relations with Arda.

Bathed, clothes washed and dried, Binda fills the pot yet again and goes back to the house. She picks clean some kudri for making into raabdi, pounds it a little and puts it on the fire to boil. After feeding Lotya, she goes out to the neighbouring field, some parts of which still have juvar stalks standing. These she cuts and uproots, heaping them together. After she has made several such heaps, she sets them alight. Thus slowly the field is cleared, made ready for the rains and ploughing. It is hot work and tiring, but then Binda works hard all day. She is used to it and accepts it as part of her lot. Her body speaks of her life of hard physical labour — she is thin but with a strength that reveals itself when she swings the axe to split a log of wood, or when she carries water up for the sixth time in the day.

As the sun sets, the juvar stalks are on fire, the flames crackling and the air smoky. Binda comes back home, pausing to drink some water, hurl a stick and virulent abuse at the hen who has climbed too close to the store of grain in the muti. The hen retreats temporarily but is soon back. Binda constantly battles the hens who run underfoot, swearing at them for going where they shouldn't, but the hens are unconcerned. I find the curses funny; literally they mean 'fuck your widow', and now, in my mind's eye, I always associate that phrase with hens clucking and scattering as Binda rushes at them. Binda checks up on the raabdi, getting glutinous now.

By the time it is dark, the raabdi is done and everyone drinks large plates-ful of it, slurping it down, then spooning up the whole grain left at the bottom with their fingers. Binda drinks some too. Then feeds Lotya. The children go out to sleep; Binda spreads her *khatla* (wooden cot with rope weaving) inside and distributes the gudris and *datis* (thin quilts and thick sheets) among all the family. She tidies things in the kitchen while listening to Khajan talk to their neighbour Dajya as they roll bidis to smoke. Khajan complains, 'Other villagers don't appreciate what I do as an activist'. He tells Binda, 'Even you don't understand why I have to travel and be away from home'. Binda jokingly tells him, 'Get yourself a second wife to help with the housework and then I will roam around with you'. Screams of enjoyment from the *kholo* draw her out to sit outside with Lotya, watching the children as they play *paani* in the moonlight. Then she goes to bed, another day done.

The Social Structure of Anjanvara: The Construction of Community

The village of Anjanvara consists of about 240 people, living in thirty-three households, all of whom are related by blood or

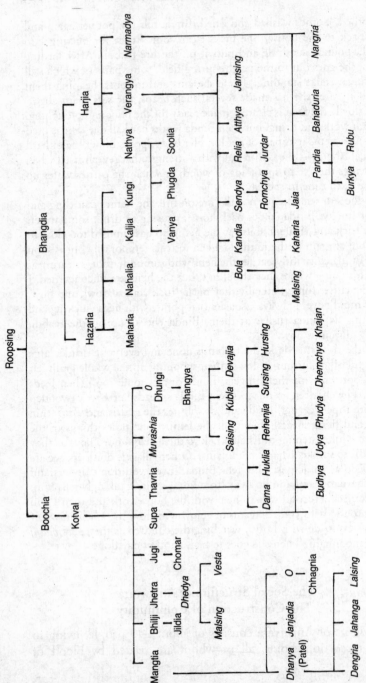

Fig. 4: The lineage of Anjanvara. Italicised names indicate living adult male members of the lineage who reside in Anjanvara. O indicates daughters of the lineage whose sons have settled in Anjanvara.

marriage. The men of the village trace their ancestry to one patriarch from whom they have descended (see Fig. 4). Their lineage belongs to the Padyarka clan, an exogamous unit.[2] The only exceptions to this rule are two men who do not belong to the clan, but whose mothers were from this village. They have been allowed to settle in Anjanvara and cultivate there by their maternal uncles. Under the circumstances, it is no surprise that most men resemble each other strongly.[3] The genealogy in Fig. 4 illustrates the relations between men, corroborating the logic of kinship on which they are based and explaining family resemblances. Just as the genealogy is a social map, so is the spatial structure of Anjanvara. The village consists of huts clustered into three *phalyas* (hamlets) strung along a shelf of land running parallel to the river. The map of the phalyas, the proximity of particular huts within them, is a telling representation of kinship, of lines of social action.

The village as a community is defined by a set of kin in relation to land. All the men of the lineage living in Anjanvara have title to, or have claims to inherit, cultivable land in the village. Only the two men mentioned earlier do not own land; they look to their uncles who have allotted them a part of their own holdings to farm. With these exceptions, the village is remarkably egalitarian in terms of land ownership; all proprietors of land are its actual cultivators. There is no landlessness or wage labour within Anjanvara. Everyone supplements their legal holdings with nevad fields in the forest. Rough estimates indicate that the legal land holdings of 'joint' families (adult brothers who have separate households but common land titles) range between ten to fifteen acres.[4] The

---

[2] Adivasis use the word *jaat* in two ways: to refer to hierarchically ranked endogamous communities, such as the Bhilala jaat, and to refer to exogamous clans within the same tribal community, such as Padyarka. Thus, the people of Anjanvara are Bhilalas belonging to the Padyarka clan. They marry Bhilalas of other clans.

[3] During the first weeks of fieldwork, I was frequently embarrassed by my tendency to confuse men with each other. Only later did I realize that my mistakes were to a large extent natural since the men really *did* look like each other.

[4] It is difficult to accurately describe the extent of land ownership and land use in Anjanvara. Land revenue records are out of date and tend not to reflect current land ownership. Records are based on *pattas* (titles to land) which are transferred upon sale or purchase of land or, most frequently, when land changes hands between generations or is partitioned among members of a joint family.

extent of nevad is even harder to gauge (see Chapter 6), but seems to be between fifteen to seventy acres. The village community also exercises the exclusive right to use uncultivated land held in common, mainly for collecting fuel and fodder.

Besides defining itself in relation to land, the community also comes together to labour. All labour-intensive tasks which exceed the capacity of the individual household are performed collectively in a form of work-sharing called laah. If there is grain to be brought from the fields to the house, a field to be harvested, or a house to be built, a family will call upon all households so that each one sends a member to its aid. In return, the family gives a feast for its helpers and, in turn, reciprocates with its own gift of labour. There is no differentiation here into landowner and labourer. Again, the system within the village comes close to realizing the ideal of co-operation usually thought of as only an abstraction, too good to be true.

Poverty persuades people in the hills to pool their resources such that the ethic of sharing is central to the maintenance of economic security. If people do not help their kin in their hour of need, who will come to their aid? This philosophy extends beyond the family in the village to the larger network of affines and agnates in other villages, constructing a web of favours and obligations in the currency of labour and kind. When a family's grain stocks run low, or its bullock falls sick, relatives usually help out. The moral imperative to share underlay the decision to let Bhangya, the present budva, settle in Anjanvara even though he

---

Unless this transfer is recorded, there is no official recognition of the people who have current rights to land. There is usually a time lapse, sometimes of several years, before the patvari (revenue official) records changes in ownership. The pace of the bureaucracy is generally slow; it becomes slower still unless a bribe is paid. For instance, in Anjanvara, until 1989, Dhanya and Shankrya's families held one land deed in common for twenty-two acres of land, even though their families had been cultivating separately for at least ten years before that. Even though land is registered as jointly held, it does not reflect further partition. According to the records, Dhanya and his three brothers are today the owners of eleven acres. Actually this land has been divided and is cultivated in three separate parts by Dhanya, his brother and their sister's son. Dhanya also has three married sons who have earmarked bits of his land for themselves. There is also an informal understanding about which remaining parts will devolve to the share of his two younger sons.

does not belong to the lineage and is only related to it through his mother. However, the idiom of sharing should not be taken only at face value; it is also a way of garnering goodwill and improving one's social standing.

Relations with Other Villages: Marriage

The agnatic lineage of Anjanvara survives over time because of its relationship with other villages. Members of the lineage sometimes settle elsewhere, in villages of their kin where they may be offered more land or some other amenity. More significantly, the community replenishes itself through marriage with people of other clans from other villages. In Anjanvara, there are thirty-five married women who come from sixteen villages. Of these, eighteen women come from villages across the river in Maharashtra and seventeen from villages on the same side of the river. Women born in Anjanvara marry outside and construct a similar web of kinship.

The many aspects of marriage — sentiment, sexuality, social reproduction, political alliance and economic transaction — are encapsulated in the events marking a wedding.

In Anjanvara, Bhika, a young man about twenty years of age, was on the lookout for a bride. At this time, the patel had guests from his daughter-in-law's family and the party included the patel's daughter-in-law's sister, Rayadi. When people started teasing Bhika about her, linking their names together, he dropped in to visit the patel's household, to sit and smoke a bidi with the patel's son. Rayadi returned to her village, but was back within a week to visit her sister, this time unaccompanied by her parents or brothers. Matters came to fruition — Rayadi liked Bhika, his gifts to her, his village, the proximity of her sister, so much so that, three days later, she eloped with Bhika.

However, the actual act of elopement fell short of my romantic expectations. This was no dramatic flight from disapproving parents, but a brief walk from the patel's house to Bhika's with the full knowledge and consent of the community. At night, Rayadi left her sister's home, escorted by some of the young girls of Anjanvara. She stopped for a while at the pujara's house on the way, where various men and children craned to catch a glimpse of her. Then she was escorted to Bhika's house and, presumably, to his bed.

Thus far, nothing had been discussed with Rayadi's family; the issue of brideprice had not been settled. So, in effect, Rayadi had run away with Bhika. Rayadi's father was informed and, two weeks later, he came to Anjanvara to demand a brideprice, failure to satisfactorily negotiate which could lead to his taking Rayadi away. The men of the village gathered in Bhika's uncle's house to bargain with Rayadi's father and his brothers. The first day, the talks went unconcluded. Bhika's family wanted to pay part of the five thousand rupee brideprice in kind and brought out two bullocks which were inspected by Rayadi's family but found wanting. The next day, after some more wrangling, it was agreed that the entire sum be paid in cash. Bhika's family put together the money with the help of contributions from other households. Then, in the presence of the assembled men, the money was counted and handed over to Rayadi's father, the transaction ceremoniously sealed with a shared drink of huru. They killed and ate a chicken to mark the event, but fixed a day twelve days later to properly celebrate the wedding with more liquor, goat-meat and dancing.

On the day of the wedding, the pujara began by worshipping the mortar-hole sunk into the ground of Bhika's house. The hole was filled with grain and two girls stood holding pestles upright in it — the explicitly phallic symbolism of the ritual and the associated imagery of female sexuality and fertility an anthropologist's delight. Then the men gathered at night in Bhika's house. Rayadi sat in the kitchen with the girls of the village. As in other Bhilala rituals, married women were markedly absent. Rayadi is very self-possessed despite her youth. She has worked as a migrant labourer in the construction industry of a town in Gujarat. The Anjanvara girls are wide-eyed with wonder that she has travelled by train. Although the women of Anjanvara do not migrate for wage labour, the women who marry into the village bring with them the experience of a diversity of work arrangements and environments.

All the men who had assembled around the pujara had their names called one by one and were made to drink from a tiny pot of huru. After all the households had been thus included, one man was made to represent the lineage. As he stood in the centre, one by one men came to give him gifts of money — around twenty rupees each — which they placed on the ground at his feet. After embracing and bowing to each other, they would ceremoniously sip more huru. Then the sacred pillar of the house, where the household goddess resides, was worshipped with an earthen lamp and more huru. In this rite, Rayadi was incorporated into the household. They came one by one to stand by the sacred pillar; Rayadi touched everyone's feet and they blessed her. By this time, the food was almost prepared outside, the coarse rice cooked and the meat done. People stayed to eat but, by common consent, there was no dancing later because someone in the village was seriously ill and it would be unseemly to sing and dance under the circumstances.

While the initial steps towards marriage are taken by the couple themselves, prompted as much by the urgings of their own minds and bodies as by the nudges of their peers, the rituals of wedding reiterate the primacy of the corporate lineage. From the negotiation of the brideprice to the final celebration, the lineage takes matters firmly into its own hands, so that the alliance becomes embedded in the ongoing structures of social reproduction of the lineage as a whole. The purchase of the bride is a transaction where two tendencies militate against each other; on the one hand, the groom's party wants to pay as little as possible to acquire a woman, an important economic asset; on the other hand, the ability to pay a high brideprice is an announcement of one's wealth, a symbol of prestige. The delicacy of such negotiations, involving as they do the honour of the village, calls for the combined skills of all men, especially elders. The rites of wedding, with their repetitive invocation of all the households in turn, reinforce the collective unity of the lineage, while the gifts of money from each household to one member of the lineage reiterate the lineage's representation of itself as a corporate entity. Even as they come together to mark the joining of a man and a woman, the men of the lineage embrace and congratulate *each other*. The bride is a passive participant in all this; her incorporation into the household is marked by her bowed submission to all the men of the lineage. Thus, the woman, whose co-operation is essential for the continuity of the lineage, is ritually devalued — a patriarchal tendency that will recur throughout her life. However, this devaluation of women and the dismissal of their strength as active beings does not go unchallenged. As we shall see later, the very values which guide a community's purchase of a bride — the pursuit of honour and the desire to get one's money's worth — are defied when women flout the authority of their husbands.

The occasion of the wedding was also an opportunity for people to talk about the forthcoming campaign of the Andolan. Rayadi's father is active in the Maharashtra part of the movement; he told the Anjanvara men about recent events related to their resistance to the dam. Khajan, an activist with the Sangath, turned the gathering into a political meeting and solicited contributions for financing the new campaign. Besides being an occasion for collective decision-making and action, the wedding further cemented the alliances on which political co-operation between villages is based; incorporating comrades into the company of kin and vice versa.

Once established, a wedding produces further structures of help, reciprocity and security. Affines are the source of economic aid in times of distress; Khajan farmed in his wife's village when his land in Anjanvara was insufficient to provide for his family. Even now, he goes to his in-laws whenever he needs extra wood or bamboo which he cannot find in his own village. The presence of relatives generally sways people's choices in other transactions. As no exchange takes place between strangers, the credential of kinship determines which budva one should consult, where one might trade, who one may trust. Thus, all the households in Bahaduria's phalya take their produce to the shop in Bhitada, the village from which their wives come, while the households in Devajia's phalya sell their stuff to the shop in Khundi, the natal village of Romka, Devajia's wife. Dhedya took his ailing daughter to the budva in Sakarja, where another of his daughters lives.

Relations other than marriage are also constantly called upon to secure aid, through loans of money, material, or simply one's name and the credit attached to a good reputation. Such credit explains the respect with which the people of Anjanvara treat their relations in the village of Attha. Much prestige adheres to the people of Attha who are looked up to as the leaders of the Padyarka clan, capable of passing judgement on the niceties of Bhilala practice, and qualified to conduct rituals incorporating lower caste people into the clan. When someone from Attha asks for assistance of Anjanvara, people are usually quite obliging. Thus Avvalsing came from Attha to Anjanvara to get a bamboo basket made and was treated with great generosity. A laah was called on his behalf and men worked jointly to weave the basket. Though Avvalsing reciprocated by buying and killing a hen, it was clearly only partial compensation.

Most of the traffic between villages is people going to and fro to visit relatives — women travelling to be with a sick sister and carrying an earthen pot of buttermilk for her; young men spruce in turbans and shirts, bearing bows and arrows, on their way to their wives' village; people going to attend religious ceremonies. Visits increase in frequency during the summer months when agricultural work is slack. While remaining conscious of the economic imperatives behind such traffic — indeed, behind a good many of the exchanges between kin — it would be highly misleading if we were to reduce all interaction to an economic calculus, both material and non-material. That is, while there is an economic component to maintaining good relations with one's kin, whether to secure wood, money, labour or status, that aspect does not exhaust or, even more important, *define* the relationship. Affection is not a fiction concealing more fundamental economic links; affective ties are valued in and of themselves; the emotional security derived from them is essential and irreplaceable. This point becomes important in light of the subsequent discussion where we deal with the tensions inherent in mutually reciprocal kin relations.

Reciprocal Kin Relations: A Mystification?

The literature on 'the moral economy' of the peasant focuses on the normative role of reciprocity in assuring subsistence. In the words of James Scott, 'the peasant is born into a society and culture that provide him with a fund of moral values, a set of concrete relationships, a pattern of expectations about the behaviour of others, and a sense of how those in his culture have proceeded to similar goals in the past' (Scott 1976: 166). Peasant life is permeated by the moral principle of reciprocity, that one should help each other, especially in times of distress. Scott describes how the village economies of South East Asia have structured forms of mutual assistance much like the Bhilala laah, where the social pressures of the community reinforce the sentiments of obligation. Thus, economic imperatives are expressed through moral injunctions.

According to Pierre Bourdieu, morality is an ideology which reflects but also obscures underlying economic objectives. Gift exchange such as laah is based on *'institutionally organized and guaranteed misrecognition'* [emphasis in original].[5] In fact, such misrecognition is perhaps the basis

of all the symbolic labour intended to transmute, by the sincere fiction of a disinterested exchange, the inevitable, and inevitably interested

---

[5] The term 'misrecognition' is much too charged for my satisfaction; it seems to regard people as somewhat stupid or dense. I prefer to use the term 'denial' which attributes much more agency and consciousness to its subjects.

relations imposed by kinship . . . into elective relations of reciprocity; in the work of reproducing established relations — through feasts, ceremonies, exchanges of gifts, visits or courtesies, and, above all, marriages — which is no less vital to the existence of the group than the reproduction of the economic bases of its existence, the labour required to conceal the function of the exchanges is as important an element as the labour needed to carry out the function (Bourdieu 1977: 171).

Bourdieu asserts that gift exchange is the only mode of labour or commodity exchange to be fully recognized in societies which have an economy 'in itself' and not 'for itself'. In a formulation which echoes that of Karl Polanyi, he says that social action takes place as if collective denial of objective economic orientations was essential to 'archaic economy'. 'The "idolatry of nature" which makes it impossible to think of nature as raw material, or, consequently, to see human activity as *labour*, i.e. as man's struggle against nature, tends, together with the systematic emphasis on the symbolic aspect of the activities and relations of production, to prevent the economy from being grasped *as* an economy, i.e. as a system governed by the laws of interested calculation, competition, or exploitation' (Bourdieu 1977: 171–2 [emphasis in original]).

In the laah, too, reciprocity is described in terms which discount its transactional aspect and emphasize its moral dimension. The idiom of gift, help and generosity is consistently used and prevails over more explicitly contractual relations. The calculations accompanying exchange are usually politely left unacknowledged; indeed, it is shocking to speak of the fairness of a laah. However, people do have a finely tuned notion of appropriate exchange: Jevanti went to her uncle's laah, but came home seething with resentment. During the feast, they had run out of chicken and Jevanti and a few others were fobbed off with plain dal. Jevanti's complaints were frowned upon by elders, but her indignation *was* considered justified.

For the most part, reciprocity, whether as laah or other forms of mutual assistance, works as a stable and successful system. As the reciprocating parties are of more or less equal standing, exchanges tend to be balanced; in the interest of guaranteeing help for oneself, people must volunteer their own services to others. Besides relative equality among smallholders bound by kinship, the small size of the community makes interaction highly personalized. In addition, the demand for labour is spread out over time and limited enough to be locally satisfied through laah.

The successful management of labour needs through laah prompted a Sangath activist to suggest that 'wasteful' expenditure on feasting be eliminated and the laah be made manifestly mutual-aid among equals. That is, people should come together to co-operate on a more explicitly contractual basis, acknowledging the economic aspect of reciprocity, without incurring the social costs of feasting. But such economic objectivism fails to address the subjective representation of laah and the centrality of the idiom of gift to the character of the exchange. It does not recognize the *necessity* of the idiom of gift. The reason why people are reluctant to abandon the structures of help and generosity in favour of the more 'sensible' straightforward relations of contract pertains to the subtle politics of the laah.

Since control over labour is but one form of wealth, the ethics of laah seek to share labour, the source of value. But within that system, there are calibrations demarcating the movement of labour. Larger families, or those which are better off, will send their young sons to other people's laah. Poorer households will be represented by their heads or adult sons, who seek to garner goodwill in this way even though they are hard-pressed to spare the labour. The laah of a better off household will be well attended. The ability to call a laah is contingent upon prestige and power which, even in this egalitarian setting, tends to vary from one household to another.

Bourdieu attributes the 'primal undifferentiatedness' of labour circulation, i.e. the collective refusal to acknowledge the calcula-tions embedded in exchanges such as laah and their conflation with the symbolism of feasting, to the opposition between the labour period — which in rain-fed agriculture is particularly short — and the production period during which crops grow and ripen. This opposition gives rise to 'one of the basic con-tradictions of that social formation and also, in consequence, to the strategies designed to overcome it' (Bourdieu 1977: 179). Each household would find it impossible to maintain continuous-ly (throughout the production period as well) the entire workforce that it needs during the labour period. In order to circumvent this obstacle, households try to build up their symbolic capital of honour, prestige and goodwill through the gift economy so that they can successfully call upon their kinsmen for help. They seek to attach labour without incurring the costs of maintaining that labour throughout the long production period. Symbolic capital, which in the form of a family's status is readily convertible back into economic capital, is perhaps the most valuable form of accumulation in a society where the severity of the climate (the major work of ploughing and harvesting has to be done in a short time) and the limited technical resources (lack of assured irrigation, harvesting with a sickle) demand collective labour.

Thus the constant maintenance of the idiom of kinship, coupled as it is with differences in symbolic capital, lends a degree of flexibility in mobilization to a highly constrained system. When the task of controlling others' labour cannot be displaced to relatively autonomous institutions such as the market, but must be done directly by people, it cannot be overt and 'must be disguised under the veil of enchanted relationships, the official model of which is presented by relations between kinsmen; in order to be socially recognized it must be misrecognized . . . ' (Bourdieu 1977: 191). The currency of obligations, trust, gifts and gratitude is the foundation of both dependence and solidarity. This is its fundamental ambiguity.

However, there are considerable checks on the accumulation of symbolic capital which prevent differences from being amplified. The relative equality of resources places limits on the extent to which one household can gain at the expense of others. Equally significantly, since symbolic capital is the collective approval bestowed on a household by the rest of the community, it is much more subject to social pressure than other forms of capital. Better off members of the group have to reckon with the interests and sentiments of the collectivity because their legitimacy is based upon it, as is their ability to mobilize. Above all, wealth implies duties. In a later section, we shall see how local politics is built around the issue of the symbolic capital of honour, the privileged mode of accumulation from which obtains economic security. This insight is an important corrective to views which, charmed by the apparent solidarity of the kin group, overlook the hidden tensions of co-operation. Kinship is not a pastoral idyll but is riven with jealousy, feuds and simmering grudges.

Having said that, a caveat is in order. The analysis of relations of production based on kinship which employ symbolic capital to secure aid, including help in the form of labour, would be guilty of reverting to economism if it dismissed or devalued these relations' normative content, their subjective meanings. While the notion of symbolic capital extends the notion of the economic to include values and action which were earlier thought to be non-rational, it does tend to commit the opposite sin of reducing action to economics, albeit an economics which encompasses a much wider set of relations. The people of Anjanvara respect the notion of reciprocity as much for its *moral* quality as for its economic logic. Sharing and helping is *good* in and of itself; if we were to reduce the moral force behind sharing to an economic base, we would be indulging in our own bit of misrecognition.

The everyday life of the community depends upon the constant invocation of the notion of community and the norm of reciprocity — ideals that run counter to the desire for accumulation at the expense of others. The ideology of community focuses on men, and structures our thoughts towards their corporate unity, directing our attention away from the women — essential to the patriarchal system yet devalued within it. In the following section, we will examine the way in which these relationships within the community give rise to a distinctive politics organized around the pursuit of honour.

The Politics of Honour

In the previous discussion, we have seen how community is defined by a set of social relations which are often contradictory. The unity of the lineage, and the web of kinship which it forms with other villages, is marked by powerful inclinations towards reciprocity *and* conflict.[6] Even in the case of laah, the idiom of mutual co-operation between equals is diluted, even undermined, with the flow of labour being determined by fine distinctions in status between households. Status depends upon the accumulation

[6] For instance, when a woman leaves her husband for another man, previously harmonious relations between her father's family and her husband's are shattered. Bitter conflicts over reparations ensue, with the husband demanding return of the brideprice that he paid, and the father usually cavilling. These disputes are primarily about economic transactions gone sour; but success in their conduct depends upon a person's ability to mobilize symbolic capital, to rally one's kin to the justice of the cause.

of symbolic capital — honour, prestige and goodwill, the pursuit of which anchors local politics.

The politics that kindles the most passion, which people find the most engrossing, occurs at the level of the village and between villages, around the issue of honour. The good name of a person or a village is a value that is cherished and defended as the source of legitimacy. At the same time, feuds over women show that honour is primarily a male concern which women may scoff at, thereby initiating a subpolitics of gender. By examining four feuds in detail, this section will discuss the frequent conflicts within the Bhilala community which arise over honour and its obverse — insult and injury.

At first glance, the issue of honour may seem like a digression, trivial in comparison with the more pressing concerns of rights to land, forest and river. Certainly, its frame of reference rarely extends beyond the adivasi community to 'development', 'the state' or 'natural resources'. But Bhilalas see the defence of honour as intrinsic to social reproduction, their continued legitimacy as a social group is contingent upon their maintenance of status. Indeed, control over material resources within the community is related to the accumulation of symbolic capital. But while honour has economic effects, its significance is not limited to them. Honour is a value in itself, which teleological explanations cannot fully account for.

The politics of honour has its problematic aspects. It is not about adivasi resistance against outside domination. A lot of it seems quite needless, prompted as it is by drunken insults which degenerate into violence. From the point of view of social justice, it champions quite the wrong causes; it is not 'politically correct' to fight for the maintenance of caste taboos, for instance. But since these are matters of overwhelming concern to the people themselves, we cannot ignore them. Not to speak of feuds would be to sanitize politics into a tidy, albeit limited, dichotomy between dominant states and markets and subordinate adivasis. The politics of honour does not fit this scheme; it is partially autonomous from external constructions of what politics should be about. Even though its incorporation means sacrificing elegance in one's theoretical framework, this level of politics must be given as much serious consideration as any other. If we are to privilege people's own understanding of what matters, then it becomes essential that feuds not be neglected in this analysis. Such local politics is crucial to a delineation of the community and its total universe of sentiments, concerns and priorities. It also provides the context in which other politics — that of the Sangath and the Andolan — take place, all mutually conditioning each other.

Feuds and the Economy of Honour

In Anjanvara, Reliya had been sick for some time. One day he was heard to rail against the people in his house, saying that evil spirits among them were responsible for his ailment. Upon which, his daughter-in-law Vopari, reading into his complaint an insinuation about herself, went back to her father's home in the neighbouring village of Khundi, complaining that her father-in-law accused her of being a daakan. Her father demanded compensation for this insult: how dare Reliya call his daughter a witch? A team of men from Anjanvara went over to Khundi to clarify the matter and were assailed by a storm of abuse. They kept their cool and asserted that Reliya's grumbling had been general; it had not been directed at Vopari. After all, Reliya had continued to eat food cooked by her; if he had thought that she was a daakan, he would never have done so. Vopari also concurred with this version of the incident. But Vopari's father had assembled a group of his fellow-villagers to negotiate. They said that Anjanvara could not get off so lightly; they would have to redress the insult to Khundi's reputation. Anjanvara's men refused to give in and went back home, the dispute unresolved. The next day, Khundi acknowledged its defeat by sending Vopari back to her husband's home.

In another episode, during an indal celebration in the village, a group of young men from Anjanvara asked one of the guests, a girl from Bhitada, to come out of the house and eat some paan. She went with them and there was some horsing around. It is unclear what really happened: they said that she spat betel juice on their face and provoked them; she said that they pulled her and fondled her breasts — but she went back and complained to her father, Kaharia. An enraged Kaharia, more than a little drunk, insulted his host, but was quietened down. The people of Anjanvara made light of the incident, dismissing it as mutual fun among spirited boys and girls.

Kaharia did not let the incident pass but brought it up when Anjanvara men, dressed in carnival costume, went to Bhitada to celebrate Holi. During the festivities, Kaharia accosted Anjanvara's patel about his daughter's molestation; but the patel drunkenly boasted that

Kaharia's daughter was nothing; if Anjanvara wanted they would have their way with *all* of Bhitada's women. This was unbearable provocation. As the revellers from Anjanvara turned to go back home, Kaharia let fly a stone. Some more men from his village joined him and pelted rocks at the departing people. Anjanvarias fled, their feathers and bells in disarray. They had shown commendable restraint by not retaliating and had prevented an escalation of violence, but they had also lost face by not standing up to such a direct challenge.

Bhitada and Anjanvara are neighbouring villages and are closely linked by marriage. In fact, Kaharia's sister is married into Anjanvara. The feud could not be allowed to fester unresolved. Two mutually acceptable *panch* [7] — a Bhilala elder and a Sangath activist — were summoned to sit as consultants and brokers to the quarrel. They camped in the village of Sirkhadi, neutral territory, where representatives from both the sides came to put forth their respective cases. During the three days of negotiation, Anjanvara was bereft of men, all of whom were sitting in a relative's house in Sirkhadi and taking part in the bargaining. Anjanvara's case was straightforward: in the presence of many witnesses, people from Bhitada had attacked them violently. Not only were they physically hurt, they had become the laughing stock of the community. They were the injured party. Kaharia of Bhitada was awkwardly placed: if he had peacefully protested against Anjanvara when his daughter was molested, he would have had the moral high ground. But by initiating violence out of proportion to the wrong done to him, he had caused injury to Anjanvara. Within Bhitada too, there was some resentment that Kaharia's impetuosity had got them all into trouble.

Anjanvara opened the negotiations by demanding that Bhitada pay them three thousand rupees as compensation. Bhitada offered five hundred rupees. Anjanvara's figure was thought to be outrageous by the panch, who managed to hammer it down to the still substantial figure of one thousand rupees. Anjanvara refused to go lower than this, and

[7] The attempt to settle disputes through the intervention of panch (elders) is a recent phenomenon which seeks to revive the adivasi tradition of peace-making with the help of a local *bhangjadya* (broker of quarrels). Usually, conflicts are taken to the police, or even to court, and both parties incur substantial expense in the process. The final judgement tends to be contingent upon the relative financial capabilities of the contestants to grease institutional palms. Villages which belong to the Sangath try to keep the flow of resources and political power within the community by using the panch wherever they can. It is also interesting that the panch do not sit in *judgement*; they are mediators who enable dialogue between the two parties. They offer their wise counsel in the process of negotiation and attempt to build consensus towards a resolution of the conflict.

the panch pressed a reluctant Bhitada to accept, on the plea that the feud would remain unresolved otherwise. Kaharia paid the bulk of the fine, with the rest of Bhitada reluctantly contributing the remainder. The triumphant men of Anjanvara decided that they would use the sum to buy a big kerosene lamp and giant copper utensils which would be used for communal feasts.

The feud cooled the previously cordial relations between the two villages. Kaharia, who had given interest-free loans to his relatives in Anjanvara, now started demanding interest on them. People in Anjanvara suddenly became doubtful of the wisdom of their actions, realizing that their obstinacy had lost them Bhitada's goodwill. They said then that they would return the compensation; it was below their dignity to accept money grudgingly given. Kaharia's sister's husband anxiously reminded everybody that the people of Bhitada were their kin and that it would be best to forgive and forget. His wife was upset because her brother Kaharia would no longer eat at her house.

The feud between Anjanvara and Bhitada simmered on through the summer, with misgivings on the part of Anjanvara and resentment on Bhitada's side. Usually affectionate relations between kin had been fractured because of the escalation of the conflict. Molestation had led to the stoning, which both parties thought ill-judged and excessive. Anjanvara cited its injured pride as reason to drive a hard bargain, and thereby further alienated Bhitada. The high price paid for the insult had restored Anjanvara's honour, but they had tried to profit monetarily from the feud; Kaharia expressed his anger at Anjanvara's greed by paying them back in their own coin and transforming his loans into profit-making economic transactions.

The conduct of a feud highlights the contradictions inherent in the relations between kin. Matrimonial alliances are supposed to extend the community to which the norm of mutual regard and co-operation applies. Paradoxically, these expectations of receiving enhanced respect from one's relatives lead to heightened sensitivity to slights and shortcomings, such that real (or imagined) insults are not allowed to pass unchallenged but are taken up and fought over. The maintenance and accumulation of the symbolic capital of honour occurs at the expense of others within the community, to whom one is bound by the ties of reciprocity. Also, though the interest at stake in the defence of honour is symbolic, it inspires actions that are very directly material. Honour is partly satisfied when monetary compensation is paid. This entails another battle between the opposing tendencies of profit-making and generosity. When Vopari's father demanded compensation for the insult to his daughter, and when Anjanvara insisted that Bhitada pay them for their lost honour, they sought to turn slights to which they had been subjected to their economic advantage, even though it meant a repudiation of kinship.

Success in feuds also requires the mobilization of community resources at the level of the village, and co-operation within the lineage. The main parties in a feud rely upon the united support of the rest of their village which lends them legitimacy. In order to mobilize, the threat to the honour of the collectivity must be convincingly projected. While the norm of village unity is invoked to attach people to one's cause, help does not come forth automatically. Disputes and distrust *within* the village can conspire to scuttle even the cause of genuinely aggrieved parties. In the feud with Bhitada, Anjanvara was united because several of its men had been injured. Kaharia, on the other hand, was somewhat isolated within Bhitada, since many people felt that he had unduly escalated the conflict.

Feuds as the Politics of Gender

In both the cases described above, women were central to the conflict, yet they were relegated to being the objects over which male honour was exercised. Vopari's father was concerned as much about the insult to *his* reputation as to his daughter's. Kaharia was upset for the sake of his daughter but also for his own name. In subsequent events, people seemed to lose sight of the original injured party — the woman who was molested. The negotiations of the dispute, as well as the resulting transactions of money, take place between men. Just as in the bargaining of brideprice, in the feud too, women are primarily treated as values which *men* protect and exchange, as assets from which they derive the benefits of production and reproduction. Injury to a woman violates her male guardians' rights and has to be avenged. As the patrilineal community, as much a creation of women as of men, mobilizes to defend its honour, it denies women their agency and identity (Spivak 1988: 28–31).

Yet women undermine that system of exchanges between men

by asserting their powers of resistance within the constraints of this system. The first years of marriage are remarkably unstable. It is common for a young woman to run away with another man after her father and husband have completed the transaction of brideprice, and set a chain of conflict into motion.

Kekdia had paid part of the brideprice to a man in Bhadal who said that he would send his daughter to live with Kekdia only when the entire sum had been paid. Meanwhile, gossip reached Anjanvara that the woman in question had started living with someone else. Kekdia immediately mobilized a party of Anjanvara men to go and impress upon the woman's father that he could not renege on the deal. They took with them the remaining amount of the brideprice to clinch the issue. Despite the father's assurances, days passed before the woman came to Anjanvara. When she did come, it was evident that it was against her will, for she left after two days. Kekdia went to Bhadal and threatened her father who sent her back. Within the week, Kekdia's wife fled yet again. By this time Kekdia had become the butt of sly jokes and quiet ridicule. Whereas earlier, his cause had been championed by the village as a matter of their combined honour, his wife's behaviour had undermined his credibility and made him look foolish.[8]

A man's honour depends upon his ability to exercise control over women. Whereas Kekdia's fight with his father-in-law was generally thought to be legitimate, support for him eroded when he was seen as being incapable of enforcing his rights over his woman. However, people in Anjanvara, including the women, were united in their condemnation of Kekdia's wife who they described as being 'of bad character'; the women of the village accepted, for the most part, the dominant values of patriarchy.

In local conflicts, women, who men treat as value par excellence, are also agents and subjects constructing their own politics. They can manipulate male notions of honour to serve their own interests by calling upon their father or husband to their aid. They can also refuse to respect the rules of male domination which emphasize that marriage is a transaction between men and which call for women's passive acceptance of the deal. By eloping, women make men into weaklings and cuckolds, objects of scorn and

[8] In consonance with the patriarchal denial of a woman's individuality, Kekdia's wife remained nameless in Anjanvara. She was only referred to as *Kekdian ladi* — Kekdia's wife — and, till the end, I could not discover her name.

mockery. However, women's acceptance of the ideology of honour tends to be ambiguous. In their youth, sexual choice can be asserted and constraining male authority defied by running away. But the freedom to resist fades swiftly as women age and bear children and develop ties through their offspring with the village into which they have married. Resistance which undermines male honour is severely circumscribed by the system of male domination under which it occurs. Women have little autonomy to decide their lives; their only freedom lies in running away from men to *other* men. A woman takes with her only her body; for even her children are the property of their father. Yet even this degree of autonomy is fought for and enlarged upon when women withdraw their labour from their husbands' households and go away with someone else.[9]

Honour and Caste Identity

Another theme in local politics is the maintenance of Bhilala identity through caste taboos about marriage. As the earlier discussion on identity showed, while Bhilalas primarily define themselves as adivasis in opposition to bazaarias, *within* adivasi ranks they assert their superiority over Bhils.

All of Anjanvara was exercised over the issue of Demchya bringing Kedli, a Bhil woman, to live with him. Demchya's brothers were mortified when the rest of the village started avoiding their house at mealtimes. There were malicious comments too — 'Khajan [Demchya's brother] puffs himself up because he is an activist with the Sangath; then how did he let this buffalo-eater[10] into his house?' Kedli's case was complicated because her father was a high Mankar and therefore fairly close to Bhilala status. However, she had previously married a Bhil whom she had left for Demchya. Her status could be regularized if Demchya and his brothers paid several thousand rupees to a clan elder to perform ceremonies for Kedli's incorporation into the clan. But before the villagers did anything, Kedli resolved the conflict by going away with a man from another village, thereby saving Demchya's family and the village from a painful and costly loss of honour.

[9] I did not hear of any cases where a man abandoned his wife. I do know of a man in Anjanvara who has two wives. When he did not get any children with his first wife, they decided that he should marry her younger sister.

[10] Bhils are looked down upon by Bhilalas because it is supposed that they eat beef, a practice that Bhilalas, like Hindus, hold to be abhorrent.

The case of Demchya and Kedli highlights a central preoccupation among Bhilalas with the maintenance of their ritual superiority *vis-à-vis* Bhils. The internalization of dominant Hindu values has led Bhilalas to zealously maintain their distance from the ritually inferior Bhils. Bhilala identity is expressed through their repudiation of Bhil customs such as eating beef. Demchya, a poor man who is getting on in years, could not find a Bhilala woman to marry him and resorted to getting a Bhil. In doing so, he faced the greatest sanction of all — ostracism by the entire village. Bhilalas define themselves through endogamy and commensality; since Demchya had lowered himself by marrying a Bhil, the village would not eat in his house. Ostracism is terrible punishment which is not only symbolic; most activities would be impossible without the continued help of the group. Much to the relief of Demchya's brothers and their families, the fate of becoming social pariahs was averted by Kedli's going away.

Feuds and Social Reproduction

Enormous energy is expended in the conduct of feuds, energy which Sangath leaders feel would be more profitably employed elsewhere. Frequent feuds also hamper political action that requires the entire Bhilala community to co-operate. However, local politics focuses on the ideology of honour, which is central to the continued social reproduction of the community. Social reproduction involves not only the biological reproduction of the group and the production of sufficient goods for its subsistence, but, inseparable from this, the reproduction of the structure and ideological relations through which the activity of production is carried on and legitimated. The honour of the village or the lineage is hotly defended, for the legitimacy of the group, in its own eyes and in the eyes of others, derives from the symbolic capital of honour. As a value central to Bhilala identity, honour is considered to be just as important for survival as material concerns about the land.

By defending its honour, a community seeks to represent itself as worthy of respect. While honour affirms collective identity, it is also an ideology that lends legitimacy to a form of domination over women and over ritually inferior groups such as Bhils. The good name of a lineage is based on its control over women and its distance from the Bhils. Honour as a form of value is simultaneously ideology and capital. It is ideology because of its role as a principle organizing political life; and capital because it is a form of accumulation which can generate returns, both symbolic and material. Inherent in the moral character of honour is the stream of economic benefits to which it gives rise. Not least of all, in the form of compensation, honour is directly convertible into money. However, there are collectively placed limits on such accumulation. As discussed earlier, honour derives from collective approval; because it is granted within a relatively egalitarian community of men, tendencies towards social differentiation are circumscribed and contained. Yet adivasi community is based on the simultaneous acknowledgement of the value of women, and the denial of their agency. When we conceive of community, then, we need to think not only of the solidarity of the village, but the way in which that solidarity is continually challenged and resisted by women.

The politics around issues of honour is always animated, eliciting even more enthusiasm than Sangath or Andolan politics. This is partly because the entire community of men is involved as active agents who have the power to act individually and collectively. The elderly, for instance, are respected as *elders*, whose wisdom qualifies them to sit in an advisory capacity as panch. Negotiations are the business of the whole group. All men enter into the decision, passionately discussing and evaluating the opposition's delegates' proposals, and directing the course that future negotiations should take. The issues are familiar; everyone knows the parties involved and can venture an opinion on the dispute. Control over the issues is not externally determined but is exercised by the community. Thus both for the meaningfulness of the values which it embodies, as well as for the participatory form that the conflict takes, local politics strikes a chord in most minds.

Finally, rather disconcertingly from a theoretical point of view, feuds have a strong element of disorder and randomness; they can never be completely reduced to a logic amenable to outside analysis. But most strikingly, feuds are entertaining; scandalous behaviour lets loose a flurry of wagging tongues and affords everyone a chance to indulge in the entirely human pastime of gossiping about others. The sense of absorbed play and pleasure which people derive from the feud is a value that is intrinsic to its appeal.

In this chapter, we saw the ways in which the identity of the village as a community is created by actions and arrangements about livelihood, labour and marriage. In turn, these interrelated bases of life are secured by invoking the notion of the honour and prestige of the community, themes which organize local politics. Our aim in highlighting the politics of honour is simply to draw attention to an important aspect of politics that is ignored in the process of fitting *all* of adivasi action into the coherent and theoretically ordered structures of political articulations with states and markets. Our understanding of adivasi culture would be badly served if we were to represent adivasi politics as devoid of the pursuit of honour which they hold so dear.

6

Economy and Ecology

The last chapter described the community that the people of Anjanvara construct together with their kin in other villages. The community is defined by a set of relations, above all by a set of *production* relations, affecting both material and ideological production. While the previous chapter concentrated on the contradictions embedded in relations *within* the community, this chapter will focus on the contradictions in the relationship between the community and nature, the ecological base which sustains Bhilala culture. It will examine the economy of Anjanvara through the annual agricultural cycle, a chronological device that enables us to observe the organization of work as well as the year-round variations in the ways in which nature is appropriated. While the rhythms of the seasons, and the annual agricultural cycle which is based upon it, convey a sense of regularity and stability, what we in fact observe is a winding down of cyclical time. With ecological deterioration, as the ground is cut away from under people's feet, production tends to become progressively unsustainable. However, unsustainability is not inherent in Bhilala practices but obtains from their history of political and economic subordination. Finally, this chapter will examine the responses to which people have resorted in order to make ends meet — increasing participation in markets and migration.

Anjanvara: The Use of Physical Space

The village of Anjanvara consists of huts clustered into three phalyas or hamlets strung along on a shelf running from east to west parallel to the river. On the southern rim of this shelf, about a hundred steep feet below, flows the Narmada on a wide, boulder-strewn bed. The phalyas are designated in relation to the flow of the water: the upstream, middle and downstream phalyas. The land surrounding the huts is 'revenue land', that is, legally owned cultivable land for which revenue is paid to the state. This narrow strip, approximately fifty-three acres of land, is the total area to which people in the village have title (GOI 1981: 186). Behind the chain of huts the land slopes up into hills covered with grasses and the occasional spindly anjan trees. Beyond this row, the hills grow taller still and are forested. The vegetation is typical of dry teak forests.[1] Tucked away among the higher slopes of the hills are the nevad fields, patches of relatively flat land where the forest has been cleared for cultivation.

The huts in which people live are large and spacious for there is no dearth of wood, the main construction material. The roof consists of sloping rows of baked clay tiles that people mould and fire themselves. The walls are giant woven screens of bamboo and wood; the floor is packed earth and cowdung. Near the entrance to the hut stands a waist-high wooden *maal* (platform) on which rest earthen pots of drinking water and a dried-gourd ladle or metal cup to drink from. Since fetching water is a major chore, the pots are used only for drinking and cooking; people

[1] In forestry classifications, a dry teak forest is described as a 'mixed dry deciduous forest with teak usually forming the major proportion of the crop on shallow porous or stiff clayey soil' (Champion and Seth 1968: 181).

go down to the river for all other activities requiring water. Most of the area inside the hut is one large room with partly partitioned areas for tethering livestock. There is no hard and fast division between these spaces; animals — goat kids, hens with many chicks in tow — frequently overrun the human space. An over-head maal — a broad shelf — runs along part of the house, above the reach of inquisitive livestock and young children. A ladder goes up to the large closed bamboo baskets on the maal in which grain is stored. All manner of things hang on the beams suspended from the rafters so that rats cannot spoil them — seed corn, clothes, peacock feathers for wearing at the time of Bhagoria, bow and arrows, fishing nets, dried chillies or groundnuts packed in teak leaf bundles. The floor of the house is uncluttered; there is a grindstone and a couple of string cots. The kitchen is partitioned off from the main room; the hearth is small, topped by a simple raised horseshoe-shaped clay wall on which sits the pot or the griddle. A few bell metal dishes lie around; there is usually another closed bamboo basket to store grain conveniently at hand.

The Farm Economy: Crops, Livestock, Forest and River

Most cultivation occurs on the plots of land adjoining the huts, but this is poor land and rainfed, so a family cannot feed itself from all that it grows here. Everyone has taken over some land high up in the hills and the produce from these nevad fields supplements what they grow on the land to which they have title. Despite its low quality, the land by the houses is still the best that people have and it is carefully tended to grow the most prized crops. This is where they plant corn, *tuvari*[2], tobacco, chillies, and vegetables like pumpkin, beans, *dumkha*[3] and *bhend*.[4] The field next to the house is a marvel of mixed cropping; any one such field may have upto fifteen different crops growing simultaneously in a state of creative confusion. This soil is best cared for, with maximum manuring and with stone bunds built along the contours of the land to stop the soil from being washed away by the monsoon rains.

The bulk of the land is devoted to the cultivation of the staples *juvar*[5] and *bajra*.[6] Yet the crop mix is highly diversified and rotated annually; besides tuvari, people also grow other legumes like *chaula*,[7] *urdi*,[8] *chana*[9] and *kultha*,[10] oilseeds such as groundnut, sesame and castor, coarse cereals like *badi*[11] and *batti*,[12] and fibre crops like *san*.[13] While higher-yielding varieties of bajra or juvar seed may occasionally be purchased to improve production, livestock bought from the local cattle markets, and iron bought for

---

[2] *Cajanus cajan*, a pulse.

[3] *Hibiscus sabdariffa var. sabdariffa*, red sorrel or roselle, a plant with a red stem and fleshy red flowers which are tart in flavour. The tender leaves and flowers are eaten fresh and dried to be used as seasoning. The small round seeds are used to extract oil and may be bartered for salt. Another variety, *H. sabdariffa var. altissima*, is grown for its fibre.

[4] *Urena lobata*.

[5] *Sorghum vulgare*.

[6] *Pennisetum typhoides*.

[7] *Vigna unguiculata*, or cowpea.

[8] *Phaseolus mungo*.

[9] *Cicer arietinum*, or chick peas.

[10] *Macrotyloma biflora*.

[11] *Setaria italica Beauv.*, or fox-tail millet.

[12] *Echinocloa colona*.

[13] *Crotalaria juncea*, or sunn hemp.

ploughshares, axes and sickles, inputs such as commercial fertilizers or pesticides are not used. Fields are composted with animal manure and the ash from burnt agricultural waste; the cultivation of legumes helps restore nitrogen to the soil. The frequent rotation of crops, intercropping, and the use of hardy local varieties makes farming less susceptible to pest attack. The fields are often bunded with stones cleared from them which prevent soil from being washed away with rainwater. By catching soil from the higher end of the field, stone bunds also help to level the land.

Much thought and care goes into the upkeep of farm animals. Bullocks are used for ploughing, planting, some weeding, threshing and carrying. They are worshipped at Divali for a farmer cannot work without draught animals, and a household without a pair of bullocks is poor indeed. Most households also have at least one milch cow. Very little milk is drunk as such; most of it is set as curds and used for making ghee for sale; only the whey is drunk. Four families in the village own buffaloes too. One night a group of young girls sat in the moonlight in one of the phalyas, singing songs. Soon after they began, the father of one of the girls came and asked them to stop because their neighbour's cow had died the day before. How could one sing in the company of such affliction?

The ownership of livestock varies between households in Anjanvara. Since every family has approximately the same amount of land, wealth is dependent on the number of cattle, goats and hens that a family owns. While every household usually owns one or two pairs of bullocks, the patel owns five pairs. The number of goats ranges from two to thirty, with most households possessing between seven to fifteen goats. The forest is an abundant source of fodder — the children who take the herds to graze every morning and afternoon can identify more than sixty different trees, the leaves of about fifteen of which can be lopped and fed to goats and cows.[14] People also store hay from their fields and grass cut from the forest to meet their fodder needs during *chaumasa*, the busy season of planting and weeding when the children cannot be spared to go herding.

Goats and hens are well-suited to the Anjanvara economy; they need very little care and they are an important source of nutrition

---

[14] Appendix 1 lists trees and their uses that children can identify.

as well as economic assets that can be traded. All religious ceremonies, vows to the gods, and grand gestures of hospitality require the killing of goats and chicken. Children keep their eyes and ears open for signs of meat to be cooked in other people's homes, so that they can nonchalantly hang around and get invited to share the meal! One day a fox killed a hen near Khajan's house but was chased off before it could drag it away. His wife, Binda, was chagrined by the loss, but the children's eyes gleamed. A happy accident; they were soon charring the feathers off on the fire.

Being by the river, the people of Anjanvara have easy access to a source of water for their cattle and for themselves. They can drink and bathe in clean water the year round (except during the monsoon); and fish caught in the river is an important source of protein in their diet in the summer months. People catch and eat more than twenty-five different kinds of fish.

The forest is the source of much more than fodder; fuel, fibre, fruit, house-building material, medicines and edible gums figure among items too numerous to list here.[15] To name only five of the most important species: bamboo is woven into baskets, some of them four feet high and equally wide to store grain; arrows and bows are made from bamboo as are fishing traps and brooms and axe handles; strips of bamboo line the roof so that baked clay tiles can rest on them. The flowers of the muhda,[16] the tree prized above all others, are eaten as well as dried and distilled for liquor; the fruit is pressed for oil. Anjan,[17] from which the village gets its name, yields a fibre that is used to make rope; its wood is burnt as well as used for making cots; its leaves are eaten by goats. Then there is *temru,*[18] its leaves are rolled to hold tobacco and smoked as bidis; its fruit eaten; its wood used to make carts. Finally, teak is used to make all agricultural implements — ploughs, hoes,

[15] See Appendix 1 for a more detailed list.

[16] *Madhuca indica,* or *mahua* (in Hindi) is of primary importance to Bhilala culture, just as it is for adivasis all over central India. In the months of March and April, the whitish fleshy flowers of the muhda fall to the ground and are collected and dried. They are used in cooking, relished as much for their sweetly pungent flavour as for their nutrients; but they are valued even more for the huru (liquor) that is distilled from them, a clear, delicately potent brew. No ritual occasion can go unmarked by the ceremonial partaking of huru. The fruit of the muhda is sold to be pressed for oil.

[17] *Hardwickia binata.*

[18] *Diospyros melanoxylon.*

rakes; drums and kitchen utensils; its wood is burnt; its large leaves are used to make packages for storing dried chillies or groundnuts.

Access to forest and river resources enables the people of Anjanvara to hold their own economically. What is classified in government terminology as 'minor forest produce' is an integral part of what they live on. Besides self-consumption needs, they trade forest produce along with some of their agricultural produce for merchandise such as cloth, jewellery, iron implements and salt. And in years when the rains fail, the forest is refuge; they cut wood for sale in the market.

The Agricultural Cycle

In the agricultural year, work rhythms are orchestrated to harmonize with the rhythms of nature; both are ritualized collectively in the many ceremonies which mark different stages of the calendar. The year begins in April–May, the month of Okhatri. Once the year starts, people cannot sleep in the afternoon; to do so is to appear indolent, and nature bestows her bounty only on those who bring it their industry as tribute. Activity is considered as much a *duty* of communal life as an economic necessity. The relationship between work and its product is mediated throughout by a treatment of nature that attributes to it consciousness and agency, qualities that are addressed through ritual.

The new year is a time of preparation for the forthcoming season of sowing; tools are repaired and bought. Around this time, people who had migrated to the plains return home for the start of work. Houses that were being built are given finishing touches to make them ready for the rains. People also collect *kadai* [19] gum and *charoli* [20] seeds from the forest for sale. The next month, Dalvalya (May–June), literally means the 'turning of branches' — the beginning of the rains and sprouting of new leaves. But before that happens, there is a period of tense waiting, wondering when the skies will darken with rain-bearing clouds. Rain is the only source of life-giving water to the fields; the tension of expectancy tinged with desperation is palpable in the villages. [21] In most households, the store of grain is running low; in bad years, people take grain on credit from the moneylender, and live on raabdi (fermented maize gruel). Not surprisingly, this is also the time of most sickness. As drought's dry breath scorches the land, people pray for rain. Although this is a matter of life and death, people bear themselves stoically for they are perpetually dogged by the fear of drought; as rainfall is highly variable, there is mild to severe drought every four years.

The first monsoon showers are followed by heavy rain in the month of Bhudgula (June–July). Harrowing and ploughing start in earnest now, with the simple single-toothed iron implements being pulled by bullocks across the fields. The furrows are sown with a bullock-pulled three-pronged seeder. The best land near the house is planted first, then the further nevad fields in the hills

[19] *Sterculia urens* yields an edible gum used in making sweets.

[20] *Buchanania lanzan*, or chironji (in Hindi).

[21] Although complete dependence on rain is the constraint on production in Anjanvara, some neighbouring villages are somewhat better situated. Bhitada has managed to draw a *paat* (a water diversion system of earthen bunds and channels, about a foot across) from a nullah to some fields. Though it is unlined and loses water through seepage, and has to be rebuilt every year, it allows them to get a second crop of wheat from their land. The people of Anjanvara have no such assured supply of water.

are sown. This is the month of the hardest work when all the day is spent in the field and people come home only to fall into an exhausted sleep. The next month of Rakhya (July–August) is spent in watching over the young crop, hoeing and weeding. Farmers speak of the joy of these moments, when the fields are green and beautiful, the crop dancing in the breeze. They say that working is a pleasure; even when everything is done they do not go home but stay behind giving final touches, their land gladdening their hearts. To ask that he watch over the young crop, the village deity Kumpalu is collectively propitiated with a goat sacrifice at this time.

In Kelya (August–September), the harvesting of maize, bajra and chaula begins. This is done with curved sickles. The produce is brought down from the fields; parties of men and women carry the sacks over their shoulders and on their heads, to be stored in baskets inside the house or to be piled up outside like a haystack. When everyone has harvested the bulk of their crop, the village celebrates Divaha by singing the gayana of *neelsa*, the song of green things (see Chapter 7). The next day, Navai is marked by collectively worshipping the new crop and offering it to the ancestral spirits; only after they have been remembered can people eat the produce of the season. Giving thanks for nature's benevolence and respecting the prior claims of the spirits is intrinsic to a view which appreciates the power of nature over humans. People are ever mindful of the religious proprieties governing production and consumption. During Navai, people worship the sacred pillar of the house where the household goddess resides, and the grinding stone and mortar-hole in which grain is transformed into food. Certain vegetables like varieties of cucumber and bhend leaves are not eaten until the gods have been propitiated.

The month of Dasra (September–October) is marked by paying respect to the temporal power. In earlier times, the king was worshipped on the day of Dussehra, festival of kings, and appeared before his subjects in an elaborate procession, a tradition that died with the abolition of the princely states. In an attenuated version of the ceremony, Bhils still gather in Mathvad, the seat of the former king and sacrifice a buffalo calf. On the occasion, the Bhilalas of Anjanvara worship their village deity with a goat and then travel to Mathvad to kill the goat and feast. Throughout this month, the work of harvesting juvar and groundnuts goes on.

October–November, the month of Katik, goes in harvesting, threshing and winnowing. Along with grain, stalks are brought down and stored as cattlefeed; as a treat, people carry home the occasional juicy stalk of juvar for young children to suck upon like sugarcane. Chillies and dumkha are dried and stored now. People eat well. Work is tinged with contentment if the crop is good. Animals are calving and kidding at this time too. Kholo pooja, the worship of the threshing round, takes place around this time as does Divali when the bullocks are worshipped.

Over the next month, Pooha (November–December), people sell chaula, groundnuts and other cash crops, carrying it to the traders. The year of my stay in Anjanvara was good for rain, crops and prices. Everyone became cash-rich, suddenly shining in bright new clothes and trinkets newly-bought from the haat. Investments involving large sums of money — buying a goat or a bullock, brideprice for a wedding — are made at this time. The work of harvesting, transporting, storing and processing continues into this month and the next, Utraan (December–January). Stocks of wood are built up for the summer in these months. People also start making huru for the forthcoming season of weddings and general relaxation.

The month of Danda (January–February) is the start of the relaxed time. People go to visit relatives in other villages and can spare more time for political activities. The month-long Sangharsh Yatra (long march) of the Andolan happened at this time and people were able to stay away from home for many days because the pressure of work had let up. This is the season for collecting gum from halai [22] trees; the bark of every halai tree in the forest is gashed so that the gum slowly oozes out. Children participate enthusiastically in collecting gum for they are usually allowed to keep the money that comes from its sale. The resinous fragrance of the slowly drying jade green lumps of gum pervades every house. The fruit of the baheda [23] are ready to be picked at this time; they are carefully collected from the forest and the kernels removed. The most important ritual of thanksgiving and prayer, indal, can occur during these days. During the month of Danda, a freshly

---

[22] *Boswellia serrata,* or the Indian olibanum tree, which yields frankincense.
[23] *Terminalia bellirica,* which is extensively used in Ayurvedic, Unani and Tibetan systems of medicine. The fruit is sold for two rupees per kilo.

cut piece of bamboo is planted at the spot where the Holi bonfire
would be burnt later. The rate at which the bamboo turns yellow,
and the direction in which it inclines, are said to forecast rainfall
and the movement of the clouds.

The beginning of the following month of Banya (February–
March) is almost entirely given over to celebration. The most
exuberant festivals of all, the week-long Bhagoria followed by the
days of Holi, are occasions for elaborate dressing up and dancing.
Everyone flocks to the haats on the days of Bhagoria in their finest
clothes, to swing through the air on country-made ferris wheels,
to drink syrups in violent colours, to eat sweets and, most impor-
tantly, to look for a spouse. Men with flutes and drums dance
through the crowds like pied pipers, picking up a trail of dancing
men. Amidst this gaiety, groups of young women laugh as they
look at groups of young men who may approach and offer to buy
them sweet betel leaves. Bhagoria (the name of the festival is
derived from 'elopement') is primarily a marriage mart. The festival of Holi follows and the men of the village dress up in strange

garb — feathers, bells, straw and goatskin, paint their bodies ashen white and their faces black, to go and visit other villages where they are welcomed with huru.

After the excesses of Holi, life quietens down. From some villages, particularly in lean years, people start to leave for jobs in the plains. Those who stay behind fish, collect forest produce, weave with bamboo, make arrows, string cots and make fishing nets. The fruit of the temru ripens now and is hoarded and relished. Work is desultory for the most part; perhaps the reason for the increased frequency of fights and feuds at this time. The last month of the year, Kalathya (March–April), so called because this is the season when the fruit of the *charoli* ripen and turn black, is spent in collecting wood and bamboo to repair and build houses, and in making tiles. As the sun reaches its highest point and heat beats down from the sky, people stay indoors or by the water's edge, fishing and weaving, making rope. Or they shelter in the shade of the muhda tree, collecting its flowers as they fall to the ground. Now the fields are cleared; the remaining stubble is uprooted, piled together and burned, and the ashes spread over the ground. All awaits the rain.

Sustainability of Production

The regularities of the agricultural cycle suggest that peasant agriculture in the hills can go on forever, varying little from year to year. Yet that impression is misleading. The economy of Anjanvara rests upon a precarious ecological base which renders future production uncertain. The reasons for the deterioration of the physical environment relate to the state's ownership of natural resources and the drastic measures that people have had to employ to survive under these harsh circumstances. In order to subsist, adivasis have to use the forest, often in ways that make sustainable use difficult. The three main causes for concern are the use of the forest for fodder, fuel and nevad.

I go with Binda to get wood. We climb the hill behind the pujara's house. She walks briskly so I keep turning my head to quickly catch the spread of the land — the river gleaming deep blue below, the tiny houses made miniature by the mountains old and weathered, the boundaries between fields made discernible by distance, the mountains turning further away, taking the river with them.

We walk by the side of the mountain, the path fringed with feathery grasses, dipping occasionally into crevices where teak trees rustle. Anjan leaflets flutter higher up. And a tree which Binda calls *halai*. Some *kambu*. The sounds of the children herding bleating goats float to us from farther away. We walk through more treed bits, the path more shaded now and soon stop in a sparse grove of teak. There is one tree that has a dead branch on top; it has two slim trunks from one root; perhaps it has been cut before. Binda swings the axe in a powerful curve to the base of one trunk. It hits with a resounding thwack that swings around the curves of the slopes. She hits with regular, rhythmic blows, strongly cutting into the trunk. The sound of flying bark and wood chips now accompanies the death blows to the wood. It is an awesome sight and I am taken aback by the ruthlessness of it, much like the night the hens were butchered. I had imagined Binda collecting wood, gathering up fallen deadwood and lopping branches off trees. But this is killing, for the living green wood is hacked away to get at the fuel above. And one less tree now stands tall on the hillside. So I think feebly of social forestry and plantations of quick-growing fuel species — lusina, casuarina and the rest.

Meanwhile, Binda has felled the tree and is now trimming off its branches with the haft of the axe. She cuts the trunk into lengths that we can carry, splitting the logs into quarters lengthwise. The wood is thick and sometimes the axe gets stuck in between. She inserts a wedge to get the log to stay open. The wood is now a pile of thin splints about six feet long. Binda spreads the moist anjan fibre that she has brought from home and lays the logs across it, then gathers up the ends and ties them across the carefully balanced pile. Which she hoists on her head, using her foot to hook the axe off the ground to her hand and then to place it on top of the pile. Then laden with newly cut wood — their annual rings now dismembered — she walks down the slope, leaving behind the debris — a half-hewn trunk too green for her use, many slim branches and leaves, a dead tree. Does she do this everyday? Not in the summer, she says, it is too hot. Teak is best, as is anjan; other wood is too wet. So we walk back; Binda weighed down with wood, I finally a spectator of the source from which my end use is obtained.

Such deforestation is part of a series of practices which tend to treat the forest as a given, a constant. Resources seem to be used without any qualms about waste; there is no attempt at conserving for the future. This exploitation would appear callous, even criminal, to people concerned about the preservation of natural resources. However, adivasis cannot be condemned for circumstances beyond their control. The forests that they use to meet their everyday needs are not pristine but have been systematically depleted by the Forest Department.[24] The impact of people's modest requirements appears devastating because the context in which it occurs has changed; the forests have been so degraded by the state that the loss of every tree has a significance that it did not have before. At the same time, people are not conservationists either; they do not use their ecological base sustainably. The reasons for this pertain to the political economy of land ownership and are discussed in the following section.

Land Encroachment in the Forest: Nevad

Most of the land cultivated by the people of Anjanvara lies high in the hills and consists of nevad fields. The word nevad, which literally means 'new field', denotes fields that encroach upon land belonging to the Forest Department. Trying to assess the exact extent of nevad cultivation is difficult. Nevad is illegal and people shy from telling outsiders about it. Even sympathetic questioners elicit only vague replies. There is also the problem of measurement.[25] Sustained agitation by the Sangath for the registration of nevad lands compelled the Forest Department to conduct a survey in 1988. According to this survey, each household cultivates between 15 to 70 acres of nevad. The total nevad holdings of 22 households in Anjanvara amounted to 480 acres. Since legal land holdings of 'joint' families (adult brothers who have separate households but common land titles) roughly range between ten to fifteen acres, the produce from nevad fields is a crucial part of the agricultural economy. The Forest Department

---

[24] Without access to Forest Department records, it is difficult to arrive at a quantitative estimate of deforestation in Alirajpur beyond that indicated indirectly by the revenue records (see Chapter 3). However, interviews with villagers, coupled with the physical evidence of the roads built by the Forest Department that criss-cross the hills, attest to the extensive logging that occurred in the Mathvad region. For estimates of deforestation in India as a whole, see Gadgil and Guha (1992).

[25] Land is measured in units that are variable and hard to convert into acres. A common measure of land is *juti*, the amount of land that can be farmed with one pair of bullocks. This amount varies with the *kind* of land that is assessed — whether irrigated, the soil type, and so on. Another measure of land is the quantity of juvar seeds that can be sown on it.

recorded in 1988 that most of the Anjanvara nevad is between 15 to 12 years old.

Aurora quotes the administrative report of Alirajpur State 1939–40 which states that about half the area of the state was under forests at that time, of which 289 square miles was reserved forest and 149 square miles was unreserved forests. Aurora goes on to say that 'the actual area under forests today, however, is anybody's guess since every year large tracts of forest lands are being denuded and added to the cultivated plots . . . The land is more suited for plantation of trees than cultivation. But a sign of acute pressure of population on land is that even the sub-marginally productive land is being brought under cultivation' (Aurora 1972: 59). What this observation fails to mention is the conflict between people's subsistence needs and the state's prior exploitation of the forest, its primary source of revenue (see Chapter 3).

State forestry, started under colonial rule, created this conflict by sharply redefining property rights. For India as a whole, by 1900, the Forest Department had taken over twenty per cent of the total land area (Guha and Gadgil 1989: 147). A system of management was instituted which reflected commercial priorities. These ranged from changing the species-mix to curtailing local use rights. As forests have always been an intrinsic part of the adivasi economy, their reservation by the state reduced the self-provisioning ability of local cultivators, and also adversely affected the management of their legally held land. State reservation of the forest increased population pressure on the remaining land. For the Bhils and Bhilalas of the hills, this exacerbated the situation, for demographic pressure in the hills had already been intensified due to their dispossession from the plains. The alienation of the forest by the state and the curtailment of local use rights reorganized production and restructured relationships at all levels between adivasis and nature and between adivasis and the state. State actions affected people in two ways. First, their rights to the forest, which had been reduced to 'concessions' granted by the state, were further reduced to the point where they almost ceased to exist. The culling of forest produce by adivasis was no longer licit, even though it continued to be an integral part of their economy.[26] Local 'secondary' use of the forest — for fuelwood,

---

[26] The official position of the Forest Department had changed recently to

construction timber, fodder, food — was deemed inimical to the productivity of the forest in terms of its 'primary' use — as a source of commercial timber. Second, people could not shift or expand their cultivation; legally, their holdings are frozen in time and space. Thus nevad, which was earlier a form of rotation that allowed people to recycle the land, became a settled, continuous depletion of a limited land base. Since legal holdings are very small, it became *necessary* to supplement them with nevad;[27] both kinds of land, together with forest produce, are just enough for people to subsist.[28]

Nevad and the Ideology of Conservation

The authority of the state has been buttressed by a new claim to legitimacy — namely the cause of conservation. According to the Forest Department, encroachment violates not only the legal

---

accept, in theory, that adivasis have a right to 'minor forest produce'. In 1989, the Madhya Pradesh government declared that adivasis were not mere labourers who did the work of collection but were the owners of non-timber forest produce (NTFP); therefore the state abolished the royalty on all NTFP except temru leaves. Collection prices were also increased. However, the new policy, which would potentially have snatched Rs 200 crores from industry, remained a still-born dream as traders lobbied vigorously against it. No formal status was given to the new policy; the 'worker to owner' order was revoked and the state government continued to impose sales tax and royalty (Sharma 1990: 181–3)

[27] The categories of 'legal' and 'illegal' are problematic too. Land settlement in Alirajpur occurred in 1949. It was a contentious issue from the very beginning. In many cases, cultivated land was registered in people's names only after they had bribed the revenue official. Fields in the mountains that were hard to reach were not registered at all. So the classification of land had lacunae that worked against adivasis, for much of the land that they cultivated was not recognized as legally theirs. Dewey (1978) has an excellect account of the shortcomings in the collection of Indian agricultural records during the colonial period, much of which is still valid.

[28] The situation of people in Anjanvara is not unusual; Chambers et al observe that, in India as a whole, over 200,000 adivasi families live in about 5000 forest villages, but possess no rights to the land that they cultivate. The Ministry of Agriculture advised the states to confer on them heritable but inalienable rights, but no progress has been achieved. 'The guiding principle still seems to be the old decision of the Central Board of Forestry that, "No occupancy or permanent right should be conferred on the forest villages" ' (Chambers et al 1989: 178).

prerogative of the state but is doubly damned for its ecological consequences. Nevad cultivators are decried as the main perpetrators of environmental damage; their wilful destruction is blamed for preventing the Forest Department from fulfilling its newfound mandate of environmental conservation. The Forest Department's efforts to reclaim its property directly conflict with the interests of the cultivators who have encroached upon forest land. When the state has attempted to seal off such disputed land, denying local access on the ground of environmental conservation, it has sparked off direct and violent confrontations — the most recent of which is discussed at length in Chapter 9.

The Forest Department's official analysis of nevad, replete with plausible half-truths, conceals as much as it reveals. First, it ignores the history — a lot of it fairly contemporary — of environmental destruction by the state. As discussed earlier, the forest was the primary source of revenue in the last century and well into the present one; most of the large-scale clear-felling was conducted under the aegis of the state. Resource depletion caused by state extraction redounded to the disadvantage of adivasi cultivators, making even limited local use of the forest ecologically hazardous. To borrow a phrase from Guha and Gadgil (1989), the removal of forests from the moral economy of provision and their insertion into the political economy of profit undermined the conditions conducive to sustainable resource use. It is also important to remember that the extent of extraction by the Forest Department and by the adivasis is quite different. While economic hardship compels people to cut trees and extend their fields, poverty as a cause of land degradation is of lesser magnitude than its obverse, namely greed and affluence. We should be careful that, in invoking the situational rationality of the adivasis' land use, we do not blame the victim by ignoring the obvious and much harsher power of the state and the market as material forces behind degradation (Watts n.d.: 10).

Second, the Forest Department's official analysis of nevad ignores the exigencies of adivasi cultivation emerging from state alienation of the forest and the more general history of commodification of which it is a part. Most legal holdings are too small to grant food security. Although people have a variety of agronomic practices to conserve soil, the constraints of farming tiny, unirrigated plots are simply overwhelming. There is acute scarcity of both land *and* capital, for that part of people's produce which is sold often fetches low returns on the market. These adversities are even more severe in years of drought. Despite being an essential part of their survival strategy, adivasi use of the forest is deemed illegal. People use an already degraded forest, cultivating friable hill soils on which sustainable production *could* potentially be obtained. But in the absence of legal right to forest land, cultivators are denied the security which would make investment in land improvement worthwhile. A farmer who lives in constant dread of eviction is hardly in a position to risk sinking money and effort into the considerable enterprise of bunding, afforestation or field levelling.[29]

There have been exceptions — anomalies to the official delegitimization of nevad. Elsewhere in India, in the Himalayan foothills, for example, cultivation in the forest is recognized and institutionalized through the practice of *taungya* — cultivators who are allowed to grow food crops in the forest provided they grow timber trees alongside. When, after a few years, the cultivator moves on to clear another patch, a forest crop has been established on the vacated ground. Taungya cultivators are thus low-cost forest workers.[30] In Mathvad, too, in 1986, the Sangath persuaded a sympathetic District Collector, who was alive to the importance of nevad in the local subsistence economy, to try a cultivation-cum-afforestation scheme on some nevad land. This fell through, mainly because of the hostility of the local administration, especially the Forest Department.

Third, the official discourse on encroachment also chooses to disregard the unofficial transactions to which it has given rise. The illegality of nevad makes encroachers liable to punitive fines and eviction. From the legal sanctions against nevad derives the power of forest officials; the rules have sown the seeds of a thriving corruption. Nevad has established an unequal yet symbiotic relationship between the lower reaches of the state and adivasis — partners in crime, as it were. The power to impose fines,

[29] The extent to which risk and uncertainty have been multiplied by the threat of displacement by the dam is another question altogether.

[30] Incidentally, I came across a community of taungya cultivators on the outskirts of Rajaji National Park in Uttar Pradesh who have been adversely affected by the recent programme of changing land use designation from forests to national parks and wildlife sanctuaries. These cultivators' rights to cut grass in the forest have now been forfeited in the interest of wildlife conservation.

confiscate produce and evict encroachers gives forest guards and revenue officials the leverage to demand and receive a steady stream of benefits — money, fowl, ghee — the 'gifts' surrendered by a reluctant and hard-pressed peasantry. At the same time, when lower and middle level officials collude to illegally sell timber from the forest to contractors, their activities go unchecked by adivasis who are rendered mute by their culpability as encroachers, for those who live in glass houses cannot afford to hurl accusations. Thus complicity in corruption continues.

In a historical context, the infringement of forest laws is seen to consist of diverse agents enacting complex and, often, conflicting agenda. Even officially, there is no one consistent line on nevad; on closer examination, the state's analysis dissolves into differing levels of understanding and acquiescence. The malleability of state structures in response to popular pressure and their appropriation at lower levels for private ends creates contradictions in the application of forest policies such that the state moves to accommodate encroachment even as edicts are issued against it. Thus, for example, the issue of 'regularization' of encroachment is a potent political issue locally and is raised every election time. Occasionally, the state *does* please its patrons and legalizes some encroachment. The Madhya Pradesh government did so in 1978. Subsequently, the Forest Conservation Act of 1980 took away the authority of state governments to permit the use of forest land for purposes other than afforestation, and the 'regularization' of encroachment had to be stopped. However, the state government often defies the directives of the law and holds out the promise of legalization. People demand and carefully preserve receipts of fines paid in the hope that their encroachment will be 'regularized' in the future; the bits of paper are their only recognized proof that they have been cultivating nevad for some time. More significantly, local collective action through the Sangath has staved off the state's attempts to reappropriate encroached land (see chapter 9). With the united strength of the Sangath behind them, people have also refused to pay bribes to appease officials who would batten off them.

Nevad and the Necessity of Resistance

As E.P. Thompson tells us, our understanding of forest 'crime'

— poaching, encroachment, wood-stealing, illicit grazing — must be seen as an assertion of customary rights traditionally held by the people and subsequently abrogated by the state as a part of changing modes of production, crimes committed by people labouring under a profound sense of injustice (Thompson 1975). In a sense, the practice of nevad conforms to this contentious terrain; the Sangath consistently argues that many of the disputed lands are part of traditional village commons taken over by the state. While each household is legally responsible for its own act of encroachment, individuals draw support from collective views and practices. A village will collectively decide whether it will extend its nevad and whether it will pay a fine. Even in the face of the dominant account of nevad, backed by the punitive authority of the state, people retain their belief in their primal right to the land, saying that 'we were here first; God placed us upon the land; we cultivate it. The government came later. Then how can it take the land away from us?' The right to decide on land use is still believed to be vested with local communities.

As much as these beliefs derive from a history of customary rights, they also emerge from very real and urgent economic exigencies. Nevad cultivation is essential to subsistence; people must necessarily encroach to keep themselves alive. Nevad fields keep the agricultural economy afloat, and even state retribution (considerably diminished now by the organization of the Sangath) has to be braved, the risk of fines borne, in order to continue cultivation. Thus resistance has a powerful underlying economic imperative. But the absence of security of tenure on nevad lands means that, in order to guarantee subsistence in the present, peasants may have to mortgage the future of the land itself. Even though nevad may result in deforestation and soil erosion, people lack the resources to do anything else.

However, the economic compulsions behind nevad operate within a social framework that values the peasant life and its relative autonomy for itself. To stay on the land, to go on cultivating — these are the modest desires of most people, their hopes for their children. The present life, precarious though it may be, is still the best, they think. In Anjanvara, Budhya who has experienced the hardship of drought describes how in lean years they would go across the river to Maharashtra:

We cut wood and sold it the next day in Shahada. We cut only dry wood, not whole trees like the Bhils; and if a forest guard came by us even he would let us go, understanding our compulsion. We got about 35 to 45 rupees each time between us — father and son [his son is eleven years old], but about half of it would get spent right there. So we could not save very much to bring back. And it is very hard work — carrying a full headload and walking in the sun all day. Then we also went over for a while to my wife's village where her brother gave us a bit of land to farm. We stayed there for two years; then it was back to selling wood.

When the alternative consists of this hand-to-mouth existence, living on the verge of starvation, we realize how much hinges on continued access to land. People's hopes and fears centre around getting good rains and a good crop; that they should not have to go to Maharashtra to carry wood on their heads. Before my stay in Anjanvara, there had been ample rainfall for two years; people had enough grain stored in their bamboo bins and their cash crops of oilseeds and pulses fetched good prices. Money earned was used to buy silver and livestock — sources of security and constancy in an environment marked by uncertainties.

Uncertainties abound, not least of all due to land degradation. The present ecological setting of the hills subjects its inhabitants to fluctuations in yield of such amplitude that, even without the claims of the state, their survival is tenuous. The cultivation of friable hill soils without accompanying conservation practices renders them vulnerable to rapid erosion. In the absence of long term data on changes in yields due to declining land productivity, our only approximate indicator of the inability of the present hill economy to be self-sustaining is the increase in off-farm employment and in seasonal migration to the plains.

Seasonal Migration of the Hill Adivasis

The hill economy constitutes the hinterland of the plains of Gujarat and Nimar. Breman, in his study of migrant labourers in south Gujarat, remarks that the hinterland serves both as the catchment basin for the irrigation of the plains and as the source of its labour: 'None of the water that is thus tapped comes back and . . . the same could be said of the surplus value of labour that is drawn from the region' (Breman 1985: 198).[31] Migration is part of a household survival strategy and 'however paradoxical this may sound, a growing percentage of landless and marginal-holding households can support themselves in the hinterland only by absenting themselves from the area for long periods. The small cultivators try to prevent a slide down the agrarian ladder or at all events to delay it, by taking part in migration' (Breman 1985: 216). While seasonal migration is undertaken as a matter of survival, earnings from which impart a degree of stability to the household in the hills, it hardly ever results in accumulation or reinvestment at home.

Historically, there has been a long tradition of adivasis migrating out of their area in search of employment. What is significant now, however, is that such migration has increased to an extent where it affects the lives of almost all adivasi households. Among other factors, Breman attributes the increase in migration to the plains to the decline of a relatively non-monetized economy. While market mechanisms have become more dominant, agricultural productivity has lagged behind in the hills. Since money must be earned by labour, and local off-farm employment has very low exchange value, seasonal migration becomes necessary as a monetary supplement to the limited production base. And this limited production base continues to shrink due to land degradation, in turn an ecological effect shaped by political economy.

Participation in Markets

The commodification of labour in the form of seasonal migration is but the most recent step in a longer history of commodification fostered by monetization. Today, it is the established practice to sell not only forest produce, but also pulses and oilseeds in order to purchase cloth, medicines, tobacco and other commodities. Occasional itinerant dealers of livestock travel through the hills looking for bargains in fowl, sometimes driving a pair of bullocks before them.[32] They are treated with distant suspicion, except by

[31] In his book, Breman describes the predicament of labourers, earlier peasants, displaced by the reservoir of the Ukai dam in south Gujarat, a grim reminder of the fate that awaits adivasis in the submergence zone of Sardar Sarovar (204ff).
[32] Among travelling salesmen and buyers who infrequently made their way

those with whom they are acquainted — then a deal may be struck in money or barter. Such traders, called *laahtia* (middlemen), move between the hill adivasis and the bazaaria traders and officials, mediating on the strength of their knowledge of and relations with both sets.[33] Yet they are seen as outsiders — adivasis who do not quite belong — by the people of the hills, socially useful but not to be completely trusted. People usually take their produce and custom to the shops in nearby villages or to the haat.

While Anjanvara does not have a shop, the neighbouring villages of Bhitada, Sakarja and Khundi do. Most people prefer to sell their produce in villages where they have relatives and some long-standing links with the trader. Besides the village shops, the weekly haats are the main markets for agricultural and forest produce, livestock and manufactured commodities. Transport is a major obstacle. In the absence of roads, carrying things on one's back to the nearest haat, more than twenty kilometres away in the case of Anjanvara, is a difficult undertaking, especially when large quantities are involved and people have to organize a laah. Carrying sacks of produce to the haat is well worth the extra effort because prices there are more responsive to the larger market. In the last year, when groundnut and pulse prices have mostly reigned high, people have profited from sales in the haat. However, since traders know of the adivasis' need for immediate cash, the lack of long-term storage facilities, and distance-created imperfections in information about price fluctuations, usually people face uncertain and, sometimes, unfavourable terms of trade for their produce. The market is for many a last resort and they value their relative self-sufficiency, taking pride in only eating what they grow themselves. With politicization by the Sangath, they are also conscious that they get cheated in the market, and are resentful

---

through Anjanvara, one sticks in my mind as the most remarkable. She was an elderly Muslim woman from the town of Chhaktala, more than thirty kilometres away, walking through the hills and buying broken old rubber slippers and plastic shoes, for which she gave boiled sweets in exchange. She sells the broken footwear to a dealer in Alirajpur and did not know to what use it was put after that. The enormous labour expended in this exercise — going from one remote village in the hills to another, collecting almost worthless bits of plastic for recycling — speaks of the superhuman efforts by which people endure poverty.

[33] An activist from the Sangath once remarked ironically that he too was a laahtia, intervening between two disparate worlds.

that the precious ghee which they make and sell for Rs 60 a kilo to the local trader, fetches more than Rs 120 in cities. In their dealings in the market, now more a necessity than before, people would like to be able to bargain from a position of strength; starting a co-operative marketing society is an oft-voiced suggestion at Sangath meetings.

As mentioned before, the haat is an arena of extra-economic exchange as well. Along with the administrative centres, it is an entrepot of adivasi–bazaaria interaction where aspects of the dominant culture are encountered. These range from Hinduizing influences to equally noticeable but perhaps less drastic changes in dress. Adivasis may take on bazaaria cultural idioms, but simultaneously transform them so that they become uniquely adivasi.

While dress may be adapted to express cultural differences, most increases in consumption seem to show the influence of the demonstration effect. Adivasis buy more commodities than they did before, including soap, tea, jaggery, shoes, kerosene and tobacco. Besides consumption, interaction with the market has also altered the social construction of time. The haat is held once a week on a fixed day in one town and on other fixed days of the week in other towns. Days, earlier unmarked, are now perceived by adivasis as constituting seven-day periods, i.e. weeks. Days of the week are remembered as days of the market such that the name for Friday is Valpur's day, Saturday is called Umrali's day, Sunday is Chhaktala's day, Monday is Alirajpur's day, and so on.

In these diverse ways, the haat forges relations not only between adivasis and bazaarias but also between adivasis and nature, affecting the way in which people think about time and space, shaping the articulation of their economy with the land and forest — what they grow and collect and how they dispose of these, their worldly goods. The move towards migration and greater involvement with markets is, in turn, a consequence of the deteriorating subsistence base of the hill economy, and people's growing inability to provide for themselves as they would like. Which leads back yet again to the fundamental inequities created by the state's alienation of natural resources.

7

'In the Belly of the River': Nature and Ideology

Earlier chapters dealt with the Bhilala community and their material dependence on their physical environment. In describing the emergence of the present from the past, my task has been to remove the fixed, taken-for-granted quality of the status quo and to render relationships problematic, introducing the elements of contradiction in the ways in which they are understood and acted upon. This chapter develops the discussion of Bhilala beliefs about nature. Of course, the separation between thought and action is an artificial one, and religious ceremonies are instrumental acts just as much as ploughing and weeding are, but my purpose here is to highlight the *theory* embedded in rituals.

Through an analysis of Bhilala agricultural rituals, their myth of creation, and their medicine, I now will explore some of the contradictions between the experience of ecological deterioration and people's beliefs about nature. While respect for nature is profound and permeates all aspects of life, it does not result in the generation of anything akin to a conservationist ethic.

Religion and the Sacralization of Nature

Contrasted with the changing natural and social world is the apparent constancy of adivasi religious beliefs centred around the natural and the supernatural. Indeed, for Bhilalas, the natural and the supernatural are sometimes indistinguishable. Hills, trees, stones are imbued with spiritual power and actively intervene in people's lives. Bablakda, for instance, is the god of trees; Aaikhada is the goddess of grass. Both have particular sites, marked by shrines, sacred to them. At the same time, the supernatural world of the spirits is also an extension of the social world of the lineage, populated as it is by the invisible presence of ancestor guardians and evil spirits (enemies of the lineage who sometimes assume the human form of daakans and possess married women).

The religious conjunction of the natural and social worlds is marked cognitively such that the village is defined with respect to a specific site and to its ancestors who inhabit that site. The specificity of the religious affinity to *this* land and no other is shown in the gayana which, in its Anjanvara version, situates people into a locality defined by the flow of the Narmada. The gayana changes every few miles so that, in the gayana of the Rathva Bhilalas around Alirajpur, there is absolutely no mention of the Narmada (Jain 1984). The sacredness of the immediate geography extends into lesser myths as well; the mountain of Ranikajal is the most important goddess of the Bhilalas; the villages of Mathvad and Dhadgam represent the slain body of Motia Bhil,[1] the adivasi chieftain who was defeated by the Rajputs. Besides thinking of land as legends writ large, particular sites are saturated with special mystic meaning. The cliffs of Veerbari near Khundi village, for instance, have at their base the eternal springs of Ambarpani where only budvas go. Here they come together to develop their powers

---

[1] The names are derived from *math* — (fore)head — and *dhad* — torso.

as medicinemen and mediators between the world of the spirits and mortals.

The ties between the village and its ancestral spirits are also spatially depicted. At the boundary of the village, usually under a tree and indicated only by special stones and terracotta figurines, stands a small shrine to Khatri Mugadya and Babu Ditya, ancestors who guard the village, to whom people pray in times of distress or at the start of difficult journeys. But it is not enough to only remember gods in moments of trouble; they must be tended in good times and bad. As the previous chapter's discussion of the agricultural cycle of the hill adivasis described it, there is a highly developed calendar of propitiation around the worship of local deities through possession and animal sacrifice. By establishing a religious connection with their environment and acknowledging its power to affect their fortunes, people also 'naturalize' their own existence, explaining and, at the same time, claiming for themselves their physical world. Appropriation, explanation and legitimation occur simultaneously. This is seen, for instance, in the gayana which, in explaining the creation of the world, also establishes a primordial relationship between Bhilalas and nature.

Environmental Action Through Religion: Indal Pooja

In analysing religious rites and myths, I am heedful of Bourdieu's stern admonition that

true rigour does not lie in an analysis which tries to push the system beyond its limits, by abusing the powers of the discourse which gives voice to the silences of practice and by exploiting the magic of the writing . . . Rites and myths which were 'acted out' in the mode of belief and fulfilled a practical function as collective instruments of symbolic action on the natural world and above all on the group, receive from learned reflection a function which is not their own but that which they have for scholars (Bourdieu 1977: 155–6).

This caution against over-interpretation is well taken. However, the performance of rituals, while certainly attempting to achieve practical effects, also enacts with great eloquence a cosmological tale — a mapping of the social and natural worlds and a theory of knowledge. This is not merely a creation of the scholar's imagination, but is manifestly an affirmation of the collectivity, in a Durkheimian sense. Durkheim saw religious belief and ritual as representing social realities in two senses: first, as a cognitive means of interpreting the social world, rendering it intelligible, albeit in a metaphorical and symbolic idiom, so that religion is a sort of mythological sociology; and second, as a way of expressing and dramatizing social realities (Lukes 1975: 292).

Such a complex effect, expressive as well as instrumental, can be seen in the collective performance of the most important Bhilala ritual — indal pooja — the worship of the union of the rain and earth which brings forth grain. I saw a smaller version of this, called kholo pooja — the worship of the smooth circle of packed earth and dung where crops are threshed and winnowed — in the month of October, the time of harvesting.[2] Kholo pooja is thanksgiving and prayer for continued prosperity. The entire village used to perform one pooja; with the branching of the lineage into separate phalyas, the pooja has also been partitioned.

While seeking to affect nature, indal also has another pragmatic purpose: that of displaying and using the wealth of the family. The more lavish the ceremony — the number of guests invited and goats sacrificed — the greater the glory adhering to its performers. Openhanded expenditure on such rituals is an act that is depicted as generous and socially redistributive; the entire community feasts at the house of the pooja performers but, at the same time, greater symbolic capital accrues to the benefit of the performers.

All the men of the village had gathered in the house of Dhedya for the pooja. The ceremony was centred around the *paatlas* (low wooden platforms) which symbolize household deities. Some earthen lamps flickered on and around the *paatlas*, alongside a few clay pots covered with *pahal*[3] leaves from the forest. A nearby heap of pearly juvar was crowned by one such pot, signifying the link between food, the forest and the household that depended upon them for sustenance. More leaves of *bel*[4] were placed on the side and on the edge of the *paatlas*. In the middle was a brass pot of huru, liquor brewed from muhda.

[2] Reverence for the earth is expressed in everyday life too: adivasis always squat on the ground when eating meals; to sit on a higher surface is to show disrespect to the earth, the giver of food.

[3] *Butea monosperma*, or *palash* (in Hindi), is a tree called 'flame of the forest' for its bright orange flowers which blossom in March. Pannia (roti-like bread) is also cooked between its leaves.

[4] *Aegle marmelos*, a tree with large green edible orange-sized fruit, valued for its medicinal properties. Hindus hold bel to be sacred to Shiva.

When everyone had gathered and all was in place, the pujara began to invoke the names of gods — the sun, moon and earth, the seas and the universe. Then he called forth one by one the heads of households in the village. As he named people, they would come forward, bow to the pujara, take the tiny *ulki* (ladle made from a hollowed out gourd) full of huru from his hand, offer a few drops to the earth and their ancestors, and then they would drink, putting the ladle to their mouths and cupping their palm around it. Turn by turn, every family was remembered and acknowledged thus, the corporate being of the village made manifest.

Then the budva slowly went into a trance. First he started shaking all over, the jerks becoming more and more violent till he was hopping around in the clearing, barking like a dog. People, who had been watching the proceedings with interest so far, became even more intent, for the budva was possessed by the spirit of God, Baapji. In the silence the pujara asked Baapji if He was satisfied, if the pooja has been done properly. The budva whispered and mumbled and everyone sat back relieved for Baapji had shown that he was gratified.

Meanwhile the hens, who had so far lain quiet under a bamboo basket, were brought out to the pujara. They came quivering and clucking in fright, to have drops of water and grains of juvar placed upon their heads. If they shook their heads, it was a sign that Baapji accepted them as an offering. Then their heads were cut off with a sickle, their fluttering bodies tossed under the basket, to be cooked and eaten by men and children in a communal meal.[5]

Then the singing started. The budva, accompanied by a small drum anchored between his feet, sang the main narrative of the gayana, while several boys sat around him, singing the refrain. As the brisk rhythm of the gayana rose higher and higher, one by one the boys became possessed and began jumping up and down, pounding the earth with their feet, rolling their heads from side to side, and barking like dogs, to later fall down insensible. The singing and the possessed dancing went on all through the night.

The Gayana as Environmental Theory

The gayana (see Appendix 2) is the most intriguing part of the indal pooja. In the context of the ceremony, as a family shares its prosperity with the community and propitiates the gods so that their benevolence may go undiminished, the gayana is not just a

---

[5] Larger poojas such as indal are marked by the sacrifice of one or more goats in addition to the chickens and hens.

story but the evocation of a set of relationships between the natural world, the people who depend upon it, and the gods who control it. The music and the text of the ritual articulate these relationships and attempt to generate an engagement with the divine.

As mentioned earlier, the poems are usually sung with a budva or a group of men leading the narration, while younger men and boys repeat the chorus. The main singers are accompanied by *dhaks* — hourglass shaped drums. The tempo of the gayana and the rhythm of the dhak rises steadily so that, during the night-long singing, men fall into a trance and are possessed by the spirit of Baapji, their God.[6] The energy released in the episode of possession transports singers and listeners into the realm of the sacred. The gayana is also sung at the time of Divaha every year.[7]

The tale of the gayana is one way in which Bhilalas explain their world. It tells of the creation of both that which is necessary for survival — like juvar — and that which is beloved — all the musical instruments. Through the gayana, Bhilalas discuss themes that are central to their universe — the creation of the natural world of beasts and men, the origin of juvar and of social differentiation. Throughout the gayana flows the Narmada, bestowing life-giving gifts to all whom she meets, naming and making sacred the geography along her banks.[8]

The story of the gayana can be divided into four parts — an act of unmaking chaos, followed by three acts of creation through unnatural births. In the very beginning was nature disordered — the mountains were misbehaving and the world was terrorized by wild creatures. Malgu gayan, the musician, played his rangai and banished the beasts from the earth. In this enterprise, he was helped by Relu *kabadi*, the woodcutter, who found wood and

---

[6] The gayana is something of a musicological oddity too. Jain observes that the *tala* (rhythm length) of the gayana is non-existent in contemporary Indian classical music. 'On this ground, as well as because some of its melodic movement is similar to what is heard in Vedic chanting, it has been suggested that this musical tradition is perhaps related to the Vedic and Buddhist chanting traditions' (Jain 1984: 3).

[7] The story in the Appendix is the main gayana; there are two stories which follow it — that of kanheri is sung at indal and that of neelsa at divaha.

[8] There is a striking parallel between this myth and those of the Australian tribes whose ancestors, in Dreamtime, roamed the country singing the world into being (Chatwin 1987).

fashioned it into the rangai. Order in nature is restored with the use of the right magical tools, wielded by the appropriate specialists. The tools themselves are acquired after the proper performance of rituals. Just as the rites of indal must follow tradition to be effective, so also in the gayana, correctly done rituals are rewarded with sought after effects. Ceremonies and music, appropriately applied, soothe nature into co-operation. That is the story of Malgu gayan.

The next episode of the gayana follows Relu kabadi's daughter Narmada on her journey to her betrothed, the sea. As she winds her way west, she grants boons and names places into being. Her many kindnesses delay her and she is jilted by her lover in favour of her sister Tapti. Fated to be forever unwed, she plants the thorny white brinjal, from which is born, along with all the pestilences, the woman artisan Veelubai. Till the end, the Narmada of the myth is benevolent and nurturing — creating, planting and irrigating, just like her counterpart in real life.

In the third part of the story, the world is shaped into being by Veelubai with the help of her daughter. But Veelubai, the master-potter, has to surrender her creations to God because of her uncommon parentage. God takes upon Himself the task of bringing the creatures to life but does not know how; His aunt and her daughters breathe life into the clay. The breathing shapes get blood in their veins when they eat things particular to their species allotted by God. This establishes the link between humans and juvar, the food basic to life's blood, the staple crop of the Bhilalas. Finally, the sun and moon are born when a mendicant in disguise impregnates two barren women. The same mendicant, with the music of his *rantha*, pleases a spirit and makes him invisible. Thus in the end, like the beginning, music is instrumental in bringing about change.

Cultural Puzzles in the Gayana

The story in the gayana has aspects that are hard to explain. Everyone in the gayana sends letters to each other. The pervasiveness of the written word — letters, studying, reading — is puzzling in a culture that is almost completely non-literate. What can explain this prevalence? Surely people so far removed from Brahminical dominance are not so impressed by the power of literacy as to incorporate it into their most potent myth? How does indal, in its entirety, relate to the dominant Hindu tradition? Jyotindra Jain interprets the cult of Ind among the Rathva-Bhilalas as 'the survival of the ancient Vedic and Puranic cult of Indra [the god of rain]. It is amazing that this old agricultural cult of Indra, described in ancient Indian literature, has survived almost intact, in a remote tribal corner, when the worship of the deity has become almost extinct from the mainstream of Hinduism' (Jain 1984: 3).

Within its regionally specific location, the gayana varies enormously. Every few villages or so, just as dress and dialect change, so does the gayana. It is a sign of the spatial specificity of the gayana that Stiglmayr's recording among the Barela-Bhilalas of Pati, approximately forty kilometres from Anjanvara where I transcribed this version, bears no resemblance whatsoever to the gayana given here (Stiglmayr 1970).[9]

[9] Stiglmayr gives two stories (pp. 146, 162) which, in some parts come close to the Anjanvara song, but there is no mention of their locale. The people of

Then there is the even more significant issue of gender in the myth. The entire indal ceremony, including the singing of the gayana, even the cooking of the feast, is an exclusively male undertaking. The woman of the house is summoned only once, very briefly, in the ceremony. While the ritual is performed by men, the world in the song they sing is created by women. Just as the absence of women is noticeable in the ceremony, the presence of women as creators is remarkable in the gayana (see Appendix 2).

Contradictions between Beliefs and Action

In the annual ritual cycle related to nature we see collective actions aimed at affecting (super)natural phenomena. Reverence for nature is combined with deep knowledge of social relations and natural processes — most of all those related to subsistence. However, at the level of belief and practice, there is little acknow-ledgement of ecological change.[10] While indal asks that capricious nature be kind, it does not address the issue of deteriorating environmental conditions. Understanding seems to be somewhat contradictory: on the one hand, Bhilala attitudes towards nature are markedly respectful, as manifested in rituals, their knowledge about their environment unrivalled. On the other hand, they act as if the natural resources at their command are undiminishing — the earth a cornucopia and the forest limitless.[11]

The belief that natural resources are for the taking, that the forest will always be there, is partly a vestige of the past when the low pressure on resources meant that regeneration could take place without any conservation measures. But today, even when people

---

Anjanvara point out that there is an *udud* (not serious, worthless) gayana about Ranikajal sung at smaller kholo poojas to induce possession. In Gujarat, there is the gayana of Pavagarh and Mataphene. In Nimar, too, the gayana is different.

[10] Of course, we can interpret the cryptic beginning of the gayana, 'our mountain is misbehaving', to be an implicit recognition of the instability of the physical environment.

[11] The novel *Paraja* eloquently describes an adivasi's attitude towards the forest: 'How vast the forest is! he would think, and how nice it would be if all these trees could be cut down and the ground completely cleared and made ready to raise our crops. Land! That is what we want and there would be enough land then' (Mohanty 1987: 22).

realize the need to conserve, the pressures of the present make foregoing for the future nearly impossible. Because there is little else that they can do, people must cut wood and cultivate thin soils. In order to guarantee present subsistence, they have to mortgage their future. As the discussion of nevad shows, this dilemma has been compounded by the history of denial of legitimate adivasi rights to the environment which has stifled the potential of indigenous institutions for resource management. The disjuncture between understanding and action remains unspanned; there is the ever-present tension of trying to carry on as usual, even as the ground is being cut from under people's feet.

Acknowledging Theoretical Inadequacy: Bhilala Medicine

In another aspect of the relationship with nature, the disjuncture between theory and practice, and the apprehension that theory is not quite efficacious, is more squarely faced. This is the case of the Bhilala understanding of sickness and disease. At first glance, it may seem that health and sickness have little to do with environmental concerns about the land. But, just like the forest, disease is considered to be a part of nature. Actions aimed at securing nature's bounty — a good harvest or good health — are based on a common conception of nature which emphasizes its supernatural aspect. Physical phenomena such as soil, water and forest, or disease, are aspects of a conscious, active and powerful divinity, whose co-operation has to be solicited. There are parallels between the ways in which people deal with the body and the land. Theories of medicine are as much postulates about the workings of the natural world as are theories of land use.

It is generally believed that evil spirits, sometimes personified as daakans (witches), cause illness. Treatment is aimed at throwing out the bad spirits lurking in the village. When three heads of cattle died one after the other in Anjanvara, people took out a sivdu. Men came carrying aloft two long poles with *moyni* bark, to which they would attach things collected from each house — eggshells, broken pot shards, old brooms — household waste. The men would enter each house, sprinkle some magical water with a bundle of neemda and bufulu leaves and grass tied like a brush, and add that household's contribution to the sivdu. The origins of this sivdu were in Mathvad where it was initiated to take away sickness; it came all the way to Anjanvara as the villages in between, one by one, took it through and threw it outside their boundary. The people of Anjanvara left it beyond the village and informed their neighbours about it so that the movement of the sivdu could continue.

In order to ensure the benevolence of guardian gods who would ward off sickness, there are regular sacrifices. If the gods are piqued because of neglect or inferior offerings, illness may ensue. The god Kuvaju was promised a particular goat which was later diverted to an indal, and, in its stead, another goat but with a broken back was offered. The god refused the sacrifice and accepted only a chicken — a sign that the god was angry. People vowed to make amends to Kuvaju the following year.

There is constant traffic in poojas and maantas, the sometimes quite extravagant ceremonies and vows made to the gods. Udya offered three hens and two coconuts to the gods for stopping his wife's stomach aches. When sickness is prevalent, the entire village comes together to perform a big pooja; every household contributes so that a goat can be sacrificed. There is always ongoing collective action to affect imperfectly apprehended natural phenomena.

The limitations of this theory of knowledge made themselves known to me when the pujara was brought down, faint with pain, from the hills where he had been working in his fields. I was summoned to his house. He said that the region around his left kidney and his stomach hurt; this was a recurring ailment. I was at a loss; I knew next to nothing about medicine; all that I had with me were some pills for diarrhoea and indigestion. What could I do? So I gave him some Digene (a brand-name antacid), ceremoniously handed from me to the budva to the pujara. It seemed to give him some relief for he was back the next day asking for more. Crestfallen, I inquired after traditional medicine: surely the pujara and budva must know how to treat a pain that was relieved upon chewing an antacid tablet? They said that they did get some medicinal plants from the forest (see Appendix 1: list of trees), but now these were found in very few, hard-to-reach spots. All through the rest of my stay, I supplied him with pills and discoursed on acidity and eating right, shaken by the awful irony that I, with my fond faith in indigenous medicine, ended up doling out Digene to a Bhilala pujara.

The Bhilala theory of sickness and healing combines negotiating with the spirit world with empirically proven herbal remedies. Diseases such as dysentery are known to be cured with locally available plants. However, this practical knowledge has been systematically devalued in the modern world. The depletion of the forest has further eroded the knowledge base of the people. While herbal medicine is harder to find because of the deteriorated resource base, that alone is not reason enough to explain the *theoretical* inadequacy of Bhilala medicine. To adopt a posture of cultural relativism and say, as Peter Winch does, that one cannot, indeed *should* not, compare the Bhilala theory of illness with that of allopathy would be to avoid the issue (Winch 1964). Both have the same objective. Bhilalas do not have a fatalistic attitude that illness and death are unavoidable companions to life, to be accepted passively.[12] They make the same efforts, usually strenuous and at great cost, to ward off and heal sickness. Most treatment consists of the budva incanting spells under his breath as he sprinkles ash from a smouldering dungcake on the body of the sick person. Sometimes, to pass on the potency of the god from his body to the medicine, the budva also spits upon the embers. Rarely does the treatment vary from this. In many cases, people find allopathic medicines more effective. There was a constant demand for skin ointments, eyedrops, and antipyretic, analgesic pills. As Charles Taylor observes, there is an inner connection between understanding how nature works and achieving technological control over it which commands everyone's attention, such that allopathy prevails over traditional medicine (Taylor 1985: 147). In terms of achieving a practical effect, it was clear that *budvai* (traditional medicine) was less successful. Under these circumstances, people in the village passed judgement on both systems by preferring one set of medical practices over another.

The greater efficacy of allopathy notwithstanding, people sometimes still prefer to stick to budvai.

Budhya's son Joiyo was moaning with pain and high fever. The joints in his body, especially one elbow and knee, were swollen, while his limbs were stick-like, his frame emaciated. The swellings on his elbow and

---

[12] However, the constant presence of death is acknowledged in the simple yet grim greeting, 'Staying alive?' asking if the children were safe, a reference to the high rates of infant mortality.

knee developed into large and angry-looking blisters. The child had not been eating; the budva said that he had fire in his stomach which made the blisters burn. The budva's treatment consisted of murmuring incantations under his breath. He also spread some crumbled dried neemda leaves around Joiyo, giving some more leaves to tie to the sores once the blisters burst. A month and a half later, Joiyo was still in bed, with red pus oozing out in a thin stream from his elbow and knee. When I remonstrated with his parents that, after a long and ineffective treatment with budvai, surely they could try allopathic medicine to save the life of their eldest son, Budhya said that they did not give bazaaria medicine for this particular disease. When I pressed him, offering to accompany them to the government doctor in Sondwa and pay all the expenses, Budhya still demurred, saying that the doctor would operate. They knew of a girl with a similar disease who died during surgery.

My frustration with Budhya's obduracy — which I thought to be his refusal to save his son's life — was, of course, unjustified. Allopathic medicine is usually inaccessible and out of people's reach; the easy availability of my small supply was an anomaly. The nearest government health centre is more than twenty kilometres away over the hills and a sick person has to be carried in a hammock — an effort involving at least four people. One of them has to be a *vaataad* (spokesman), who can talk to the bazaarias in Hindi. If the visit to the clinic involves hospitalization or has to be repeated, people have to worry about getting a place to stay and cook. Government doctors are sometimes crooked and demand money, or send people away after giving them an injection of distilled water; quacks are even worse. Most bazaarias are rude and push adivasis around. The powerlessness induced by the entire effort is part of the reason why people hesitate before going to a government dispensary.

When one considers that their encounters with allopathic medicine have been alienating in the extreme, then it is not surprising that they cling to their own, even as they employ those aspects of the modern that are accessible to them. Belief is a mix of pragmatism — taking recourse to whatever works — and faith in divine providence. The issue of disease and healing is as much that of knowledge as of *control* over knowledge. Allopathy is alien and out of reach. While the ministrations of the budva may be less effective than the skills of the government doctor in making people well, they are still preferred because they are accessible.

While the course of medical action is shaped by perceptions about access and control, it is significant that adivasis judge allopathy to be a more effective theory of health than budvai, acknowledging the shortcomings of the latter.

While individual and collective action to affect nature continues, it occurs in an increasingly uncertain environment marked by ecological and political changes. Uncertainty is experienced as declining potency and the erosion of magical powers. People say that their ancestors were *jaankaar* — learned and powerful. Now magical powers have been lost because people have forgotten how to listen and learn. While budvas are still revealed their calling in a dream, their potential made known to others through early signs such as the ability to fall into a trance, though they still learn from each other, they feel that their powers are drying up. Magic does not work as well as it used to and Bhilalas feel that loss of power acutely, remembering tales about their jaankaar ancestors of the past.

The Bhilala theory of sickness and healing, based as it is upon the propitiation of spirits, parallels their understanding of the relationship between adivasis and the agricultural-forest economy. Both express the reverence that people feel for the power which controls their life and which is imperfectly known. Yet both have failings which make them incapable of dealing with the problems of the present. In the context of their marginalization, Bhilala knowledge about nature, however detailed, does not equip them to successfully cope with their predicament of survival on a deteriorating environmental base.

In this chapter, we saw that people attribute agency and consciousness to their physical environment which is personified as a group of supernatural entities affecting the living community of the village. This understanding is the axiomatic basis of a set of theories, embodied in rituals, which are applied to deal with the phenomena of disease and fertility of the land. While reverence for nature is evident in the myths and many ceremonies which attempt to secure nature's co-operation, that ideology does not translate into a conservationist ethic or a set of ecologically sustainable practices. People's action is constrained by their political and economic marginalization, and increasing ecological uncertainties make their lives precarious. Under these circumstances, ecological sustainability is not even a concern that people can afford to entertain. All that they hope for is to go on as before, their children growing up and their land getting good crops, their low-level environmental use continuing into the future. These modest actions and aspirations, shaped in the context of overwhelming odds, must be kept in mind when we go on to discuss the claims made about adivasis by environmental activists keen to incorporate them into a discourse about environmental movements and sustainable development.

8

The Politics of the Sangath

The earlier discussion of the community (Chapter 5) described the politics within the community around the issue of honour. The pursuit and defence of honour is an absorbing social drama, an enactment of contested values and sentiments where everyone can play a lead role. However, another level of politics seeks to divert adivasis away from the infighting of the feud to channel their militancy against outside oppressors — the state and the market. This politics deals with the issue of adivasi rights to the forest and land. Unlike the politics of honour, Sangath politics has been initiated and organized by a small group of 'outside' activists. I now examine the activists' analysis of the conflict between hill adivasis and the state over natural resources, and their efforts to transform adivasi consciousness for collective resistance. I shall describe the effects of joining the Sangath at the village level where people have been able to temporarily stave off the state and thus stabilize their nevad land holdings. However, this victory owes a great deal to the resources provided by the activists. A persistent question about the engendering of self-reliant struggle against the state exercises the minds of activists and shapes their efforts. Another dilemma relates to the strategic choices that the Sangath must make in dividing its energies between the provision of welfare services and political mobilization. This leads to an examination of the activists' analysis of their historical role *vis-à-vis* the state and the people.

The Origins of the Sangath

Ten years ago, two young men met at the Social Work and Research Centre at Tilonia in Rajasthan, well known for its pioneering work in grassroots development through the organization of handicraft co-operatives, the use of appropriate technology and improvements in community health and education. While appreciating the gains of the community development model of social work, the men felt that engendering social change required direct political action through collective organization and mobilization of the oppressed against the state and the market. With this objective in mind, they came to Alirajpur in Madhya Pradesh. After the first rocky months, spent in trying to establish a base among the suspicious adivasis, they learnt of a contractor who embezzled government money by employing local labourers for wages below the legal minimum. Posing as labourers, they joined the project and successfully organized a strike demanding full wages. From that beginning, they went on to establish a base in the village of Attha where they organized people against corrupt government officials. The first stirrings of protest against the corrupt demands of the Forest Department resulted in one of the men being brutally beaten up by a group of *nakedar*s and forest guards. This incident greatly touched the hearts and minds of the villagers who felt that someone who was willing to stand by them, regardless of the consequences, deserved their support. This was the start of the Khedut Mazdoor Chetna Sangath — the Peasants and Workers Consciousness Union. At present, the Sangath consists of about ninety-five member-villages in south and east Alirajpur. While most work for the Sangath is performed voluntarily by villagers who combine activism with farming, the Sangath is led by four full-time adivasi activists who are paid modest salaries. In addition, the leadership of the Sangath includes six full-time activists, who are bazaarias.

The Sangath's Political Analysis

The most important conflict between the Sangath and the state has been over control of the forest. As discussed before, the use of the forest is intrinsic to the hill adivasi economy, and nevad holdings are a crucial supplement to cultivated land. In addition, adivasis rely on the forest for fuel, fodder, construction timber, medicine, and much more. However, the use of the forest for fulfilling these basic requirements is deemed to be illegal by the Forest Department. Cultivators can be evicted from their nevad fields in the forest; their produce can be confiscated; they are let off only after paying a fine. For adivasis, losing nevad is the ultimate calamity; the petty officials of the Forest Department have the power of life and death over them. This power has been consistently abused by officials to extract a steady stream of private profit from hard-pressed peasants. Adivasis live in constant dread of officials who make surprise raids into villages.

Before we joined the Sangath, when bazaarias walked into our village, we would run away, quickly hiding our stock of bamboo on the *maal* [overhead shelf], making sure that they could not see anything for which we could be fined. They would stop at the patel's house and order us to give them six, seven chickens at a time. They would eat them with pannia [roti cooked between leaves because Hindus consider adivasis to be untouchables and do not eat in their dishes]. Then they would point to anything they liked — a new string cot, a basket — and take it away. They would say, 'I see that your child is eating groundnuts; my children would like some too', and we would have to give them a bag of groundnuts. Or til, or chaula, or ghee, whatever they wanted. While they sat, they would run their eyes over our houses; if they saw a new house, or repairs to an old one, they would curse us for stealing from the forest. Then they would ask how much we had planted in the nevad fields in the forest. And make us pay fines. We could not protest; a word out of us and they would swing their lathis. If you dared to look them in the face, they said, *'Kyon be, bhenchod, mera lund dekh raha hai?'* [Why, sisterfucker, are you looking at my prick?] and send you flying with one blow of their rifle butt.

Before their organization into the Sangath, adivasis were so cowed down by the brutal authority of bazaarias that they had grown accustomed to the reluctant surrender of their assets, and regarded a bribe as an illegitimate but unavoidable overhead cost. The state was too powerful to be resisted. At the same time, the constant transactions between the forest officials and the people made it apparent to the adivasis that the intention of fines was not to *stop* nevad but, on the contrary, to make it possible by appeasing the state. Once the forest guard got his cut, he would turn a blind eye on illegal activities. In that sense, people look upon fines not as punitive, but as enabling payments.[1]

The work of the activists concentrated upon convincing people long inured to bullying government officials that the unofficial demands for 'gifts', solicited and unsolicited, were completely illegitimate and could be resisted collectively. If adivasis were to throw off their awe and fear of bazaarias and unitedly stand up to them, they would no longer have to relinquish their modest but precious possessions. This leap from submission to defiance could be made if people realized the justice of their cause and the possibility of successful collective action.

From protesting against corruption, the activists' political critique went on to challenge the very legitimacy of state action against nevad. This analysis was based on the history of adivasi alienation from the forest, and the denial of their right to subsistence by the government. Activists argue that, instead of being seen as a forest crime, nevad should be understood as both an assertion of customary ownership by the community and an economic compulsion. This is summed up in the slogan: *Jangal jamin kunin se? Amri se! Amri se!* (Whose is the forest land? It is ours! It is ours!) The state's reservation of the forests for commercial use violates

---

[1] Even now, fines are seen as enabling fees. Since joining the Sangath, villages have not paid fines but have also abided by their collective decision to not extend nevad. Last year, many of them decided to pay up their fines, reasoning that since they intended to expand their nevad area, they should give the state its due. As mentioned earlier, receipts of fines are also looked upon as the state's acknowledgement of land use, worth preserving as evidence that may come in handy if encroachment is 'regularized' later.

the primal right of adivasis to control the resources on which they depend. In addition, the state's pre-emption did not compensate the adivasis for their loss; not only have they not been granted alternative sources of sustenance, they have been denied even the minimal consolation of welfare services. Under these circumstances, the onerous demands of the state (legal *and* illegal) are completely unjust. If the state does not provide health care and education, if the patvari endlessly delays the recording of land transfers, then it has lost the moral authority to punish.

While concentrating on the particular face of exploitation encountered by adivasis in the hills, the Sangath activists have also tried to link their politics with that of other oppressed groups in India. The Bhils and Bhilalas who fight against nevad share common cause with adivasis elsewhere, with low-caste landless labourers and small peasants, and with the urban working class. As the name of the Sangath denotes, its members share an identity as peasants and labourers as much as adivasis. These connections are ideological as well as strategic; the Sangath joins hands with other progressive groups because they share the experience of exploitation and an understanding of its causes, and also because they recognize that their strength derives from unity.

The Transformation of Consciousness

For political mobilization, Sangath activists found that locally, nevad has been such a potent issue that it provided an instant rallying point for people. However, while relative social equality, the unity of the lineages, and local traditions of reciprocity were strengths upon which the activists could build, they had to divert the energies that were spent in the politics of honour into the collective defence of nevad. In order to mobilize people to successfully organize and assert their rights against the state, the activists of the Sangath had to contend with the divisive politics of the feud. This entailed erasing Bhil and Bhilala differences and, in their stead, fostering the development of an 'imagined community' — a nation of adivasis, joined by feelings of kinship with adivasis elsewhere (Anderson 1983: 15). Activists had to deal with Bhilala caste barriers that prohibited mingling with the Bhils. Caste proscriptions were being overlooked in public places like the haat anyway (see chapter 4); by eating in Bhil homes and by stressing the irrelevance, in fact *disadvantage,* of observing caste taboos, activists try to convince adivasis about the benefits of transcending Bhil and Bhilala divisions. This is succinctly expressed in the slogan, *Adivasi ekta zindabad!* (Long live adivasi unity!)

Adivasis in the hills had always tried to maintain their distance from the dominant culture of the bazaaria trader and government official. As discussed before, hill adivasis emphatically rejected the values and the lifestyle of the bazaaria, even as they were constrained by the condition of their subordination. The Other that dominated in the shaping of an adivasi Self was the bazaaria. The Sangath has tried to reinforce this distinction by emphasizing cultural differences and by incorporating into the disparate notions of adivasi and bazaaria ideas of class conflict, of the difference between exploited and exploiter. Their shared identity as adivasis enables the people in the Sangath to collectively oppose a common exploiter, the state. In order to do so, they have had to overcome their dread of the power of the state. For so long had adivasis been held in thrall by officialdom — from the awful *motla sahib* (big man) in his jeep whose language they did not speak, to the cryptic, ominous-looking, written records of the patvari, backed by the courts, jail and the brute force of the policeman's lathi — that they would shrink from a confrontation. The Sangath had to reinforce that part of the adivasi consciousness that resisted state hegemony, by demystifying the workings of the state apparatus, rendering its structure intelligible and therefore vulnerable.

In order to diminish the dread of the bazaarias by inspiring proud confidence in adivasi identity, activists have tried to clothe the political analysis of the Sangath in the flesh of indigenous idioms. Activists have creatively adapted traditional modes of political organization and negotiation to the circumstances of battle against the modern state. In doing so, they have also endorsed local symbols and rituals which celebrate adivasi life. For instance, the figure of the chieftain Motia Bhil who died fighting against Rajput invaders, is iconic of adivasi resistance and is still worshipped during the festival of Dussehra. By worshipping Motia Bhil during the *jangal mela* (the annual fête organized by the Sangath), activists have highlighted his stature in the pantheon of folk heroes. The Sangath has also touched a chord in the minds of villagers by encouraging the telling of stories about powerful ancestors, whose memorial stones are often still standing. This history has been woven into songs that are sung during meetings. Bamnia Naik of Jhandana, who fought for the queens of Dahi and Mathvad and was given twelve villages as his estate, and who was so famed for his bravery that when he came to Alirajpur, the king himself descended from his throne and offered his seat to Bamnia, is remembered in a song: 'Bamnia, the state has stolen your memorial; how can you stand mute?' The same song goes on to remember Chittu Bhil of Sorwa who died fighting against the British and whose captured manor is today a police station: 'Chittu, the state has stolen your house; how can you stand mute?' Singing these songs recalls past glories and reminds adivasis that their traditions of struggle and courage are being continued into the present. The songs and the legends also spark off discussions about the way adivasis used to live before and the way they live now, analysing their present predicament and the lines of action open before them. Sometimes the songs are sung with such spirit that they succeed in summoning the ancestors themselves. One night in Kakrana, about two hundred people were singing 'Bamnia, the state has stolen your memorial; how can you stand mute?' when there was a piercing shriek from the centre of the group. The body of Manglia's brother was heaving and shaking, and he announced to the hushed company, 'I am Bamnia Naik. You have called me through your song. I am pleased.' Between gasps of breath, Bamnia described how he fought in the old days, the men that he commanded, and battles that he had won. He said that he would go on watching over his people; their enemies would be defeated and they would be victorious. Then, as suddenly as he had come, Bamnia disappeared and Manglia's brother's body became limp.[2]

In keeping with the structure of adivasi communities, people

---

[2] Perhaps possession, an integral part of Bhilala worship, occurred during a Sangath meeting on this occasion because the meeting was held during the highly charged days when the mata was passing through the area. Khemla, the president of the Sangath, was also possessed by the mata once. In a trance, with incense-sticks in one hand and the Sangath flag in the other, and followed by the people of his hamlet, Khemla went to the hillock where the village goddess resides. There, he spoke about nevad: *Jangal jamin kunin se? Kalka mata kunin se? Boyda khudra kunin se?* (Whose is the forest land? Who does mother earth belong to? Who do the mountains and the streams belong to?) The mata also predicted that the police and the vania would be ruined, and that the poor would rule the world.

have become members of the Sangath not individually, but as entire lineages or villages. Of course, certain men tend to be more active than others, but by and large, villages or phalya participate in unison in Sangath activities. The kinship ties between villages have also facilitated mobilization. Thus the feud between Anjanvara and Bhitada, both member-villages, alarmed activists, who were worried that it might affect Sangath unity. The collective identity of the union is emphasized through the exchange of a greeting special to the Sangath; when people of different villages belonging to the union meet, each raises a clenched fist and calls out a hearty *zindabad!* (long live!), a shortened version of the slogan *inqalab zindabad!* (long live the revolution!) All Sangath meetings begin with a long series of zestful zindabads as people trickle in and greet each other.[3]

[3] Few women attend these meetings. The constraints of domestic work allow women little leeway for travelling to participate in Sangath activities. But the three women activists of the Sangath have tried to organize women, holding meetings in individual villages on issues such as the unequal division of labour within the household. Though they are not free to attend political meetings

Also noticeable in the strategies to transform consciousness is the Sangath's use of adivasi music. A song usually consists of a short melody, endlessly and monotonously repeated. The interest of the song lies not in the melody, but in the lyrics which are improvised on the spot, the singers taking pleasure in capping each others' efforts with more risque lines. The songs are usually made up as people dance with their arms linked, holding each other close, moving back and forth in two arcs that face each other. As they advance and retreat, one party sings out a couplet to the other, which sings back its reply. The improvised words of these songs makes them ideal for adaptation by the Sangath. People love to sing, and delight in making up verse after verse that speaks of politics instead of the usual romance or seduction.

The activists of the Sangath have also incorporated the consensual negotiating strategies of the feud into their dealings with government officials. Earlier, when a rally reached the administrator's office, one or two representatives would go in to present their demands before the official, and would come out and announce the result of their talks to everybody waiting outside. While this is generally the accepted mode of collective bargaining, it is alien to the adivasi way of conflict resolution where the panch act only as advisors and intermediaries between two parties, carrying messages and offers back and forth, and giving their opinion. Now, Sangath representatives who go in to talk to the administration do not conclude negotiations themselves; instead, they carry the administration's response back to the people who then think about it and discuss it among themselves in a way which allows all points of view to be considered. Then the response, collectively arrived at, is taken back inside to the official. This style of negotiation, borrowed from the tradition of the feud, is felt to be more satisfactory by the people.

In engendering changes in people's understanding of their situation, activists have had to contend with tremendous inertia, caused as much by the fissures within the community, as by people's lack of confidence in their ability to confront a powerful state. Through a persuasive mix of authority and sensitivity to local cultural idioms, they have managed to create a union that

---

outside the village, on the familiar terrain of their own village women have put up spirited resistance against the state (see Chapter 5).

has won for its members a stay on state exploitation, and which has opened up new horizons for collective action. The achievements of the Sangath were dramatically displayed during the episode that has come to be known as the Kiti firing.

The Sangath in Action: The Kiti Firing

Every now and then, the Forest Department decides to pursue its mandate of conservation seriously; it starts a fresh offensive to evict nevad encroachers and to exclude their animals from the forest by digging 'Cattle Proof Trenches' (or CPTs as they are commonly called) which cordon off parts of the forest. The justification for enclosure is to allow regeneration of the forest through tree plantation and protection. If guards on watch catch livestock grazing in the protected area, they confiscate the animals and their owners have to pay a heavy fine to get them back. Despite these measures, the survival rate of planted trees in most areas is so abysmal that it is difficult to discern the difference between the degraded forest and the planted portion, supposedly preserved by the CPT.

The digging of CPTs imposes immense hardship on people who are denied access to forest resources. Its unilaterality leaves no space for consultation or for the accommodation of local needs. For these reasons, the digging of CPTs is opposed by the Sangath. Thus, when the villages of Kiti, Keldi and Vakner heard that their forest was to be made off-bounds, men and women went to the Divisional Forest Officer (DFO) and laid their predicament before him. The officer was unsympathetic; he threatened that if the villagers tried to prevent the digging of the CPT, he would not let the rain fall that year (*sic*). The people replied that they would block CPT work; in response, the DFO filed pre-emptive cases against forty of them. Because of the local opposition to the CPT, the Forest Department had to bring workers from the village of Sakdi. When they started digging, the Sangath marched in procession to the work site. Men and women lay down in the freshly dug trench, getting in the way of the spades, pick-axes and shovels. Others peppered the workers with stones and chased them away. After stopping the CPT, people marched triumphantly to the administrative headquarters at Bakhatgarh, shouting slogans: *Jangal jamin kunin se? Amri se! Amri se!*

Two days later, the police combed the villages for Sangath activists. Two people were taken away from the Alirajpur bus station. One woman was picked up at night from a village and detained at the Sondwa police station; a male activist was also arrested from another village at the same time. Two of the activists were in the house at Attha, chatting late at night with friends from the village when the police walked in and arrested everybody.

I did not know about these events; I was celebrating Bhagoria at the Valpur haat, dazed by a riot of colours — men in bright turbans with coloured tassels hanging from their kerchieves, women with tattoo-stippled faces, laden with silver and bead jewellery, carrying elaborately embroidered bags. The haat was in full swing — stalls set up selling sweets, coconuts, coloured powders for throwing at people, coloured drinks, sweet potatoes and oranges. Wooden ferris wheels turned against the sky, transporting squealing girls in vivid magenta and black through the air. Men played the flute and beat giant drums and everyone danced. It was hot and noisy; the crowd had swollen to jostling size; and it was getting a little overwhelming. Just then, someone brought us news of the arrests.

Stunned, we caught the first bus to Alirajpur. The next four days were tense and anxious, spent in buses travelling to and from the district jail at Jhabua, waiting for hours to be granted permission to talk for five brief minutes through prison bars to the six activists inside. We learnt from them that four people had been arrested and sentenced illegally; they had not been produced before a magistrate, and one woman was arrested without female police being on hand. It was decided during those hastily whispered conversations that I would go to the High Court in Indore to file a *habeas corpus* petition against the Alirajpur administration for illegal detention. The only activist who remained free would inform and consult people in other villages about the arrests and perhaps plan a demonstration in Alirajpur. Meanwhile the activists in jail would go on a hunger strike. As instructed, I left for Indore by bus and contacted a lawyer-friend of the Sangath who filed a petition in court in my name. Indore also has the Sarvodaya Press Service, run by a journalist who supplies news agencies and newspapers with news about NGOs and social movements; he sent off a press-release about these events. However, even before these took effect, the threat of the hunger strike, and fear of the adverse publicity that it would bring, resulted in the precipitate release of the activists from jail. I went back to Alirajpur and the blessed relief of seeing my friends free.

While the activists were in jail — they were detained for four days — the Forest Department, thinking themselves rid of the trouble-makers, brought workers in again and started digging. The villagers of Kiti, Keldi and Vakner were holding a meeting on a hill above the CPT when two policemen came and said that the DFO wanted to talk to them. It is unclear what happened; people say that when they went down, the police and forest guards came running and rained lathi-blows on them. The adivasis ran back up the hill and retaliated with stones. Then the police fired with their rifles, seven times. But the villagers stood their ground and kept stoning the officials and the workers till they ran away. We learned of this four days later.

Then a meeting was held in Alirajpur, attended by the activists and some of the villagers. It was decided that a case would be filed against the policeman and forest guard by the men and woman who had been injured by the lathis. As the police had framed charges against the villagers again, these would be countercases to present the people's point of view before the court. Two activists went off to Jhabua to meet the District Collector about negotiating a peaceful settlement to the dispute.

For the activists, the piling up of flimsy, often false, police cases against them is doubly worrying. Not only are they harassed and their work hampered by frequent appearances in court, they run the crippling risk of being declared *persona non grata*, and being completely exiled from the district. Thus, they are trapped in a litigious mess that they can do little to avoid. Another peril is that if confrontations with the Sangath increasingly take a violent turn, the organization can be declared Naxalite, and its activists can be denied basic citizens' rights and be jailed without recourse to judicial safeguards. If the Sangath acquires a violent image, police repression can be justifiably stepped up. The move towards greater militancy would jeopardize the chances of survival of the Sangath. Thus the activists were trying to convince the people that it would be judicious to remain non-violent.

Discussions with the Collector were inconclusive. The Sangath activists were conciliatory and offered to work with the Forest Department to find alternative, less-contested tree plantation sites.

The Collector remained firm about stopping nevad, but said that some of the cases filed against villagers may be removed. He later visited the CPT site and was met by villagers who reiterated their resolve to oppose the digging. Thus the matter of the CPT petered out, the conflict between the Forest Department and the villagers at a stalemate. It seems that the administration had decided to hold its fire and let the moment of confrontation pass. Over the next months, senior officials made frequent trips to the villages to hold meetings with the people, some of whom openly mocked them. A substantial sum of money was also sanctioned for Sondwa Block to step up 'developmental' work. The people were satisfied, relieved that they had managed to secure a measure of protection for their forest use.

The Sangath in Anjanvara

For the people, organization into the Sangath has given a new lease of life to nevad. Whereas earlier, the state would have no compunction about evicting them from the forest, now it must contend with their collective opposition. Popular resistance has rendered the task of fining and eviction difficult and, for the last ten years, villages in the Sangath have lived in comparative calm, free from government harassment. Access to the forest has allowed them a degree of affluence, however temporary. Since they cultivate nevad land, they have a greater chance of growing enough to feed themselves; their life in the hills, which they value, has some stability.

In Anjanvara, the best evidence of the presence of the Sangath is the absence of the state. During my stay, *no* government official came to the village to harass the people. In fact, the village was visited by the police only once, and that because they wanted to see me.

I was in Dajya's house, splitting and trimming bamboo strips for the basket that he was teaching me to weave. As I sat whittling away at the bamboo with a sickle, Dajya worked with a chisel, making one of the wooden legs of a cot. Suddenly he said, 'The *nakedar* is coming!' and ran to stash the wood away. Curious, I went out and found a police officer with two rifle-toting juniors in tow. He said almost obsequiously, 'Madam, we have a little business with you'. His next words showed his anxiety to clarify at the very outset the peaceful purpose of his mission:

'We have brought a letter for you from your colleague. Please be so good as to read it and give us your reply.'

As they stood, a crowd of villagers gathered behind me, quiet but watchful. They later told me that they recognized the officer as the head constable of the Sondwa station. He introduced himself to me and asked if he might sit down; they had been walking all day in the sun. Someone brought out a cot and the policemen sank down on it, mopping their faces. No one offered them any water, but just stood and watched, the atmosphere tense. The officer was nervously affable and kept repeating that he had only come to give me the letter. I read the letter — a request made by the Deputy Collector through an intermediary that I come to Alirajpur to talk. It offered to withdraw the cases against the people in the Kiti episode if I withdrew my *habeas corpus* petition from the High Court. I quickly explained the contents of the letter to the people who were standing guard, as it were, and they relaxed, reassured that there was no trouble. The policemen still seemed to feel threatened and uncomfortably out of their element, for they left as soon as they could.

Once they were gone, everyone spoke all at once, Dhedya describing how the policemen first stopped at his house to ask directions. They called him *dada* (elder brother), and begged him to give them a drink of water! Other people said that they first thought that I would be arrested, but knew that the police would be too scared to do it in their presence.[4] From other villagers I learnt that the police had started out from Attha and had lost their way several times. People would not even give them water to drink. At one place, they had begged for food and were given cold *kulthyan ghugri* (a bland, stodgy mess) to eat. Quite a change from the days of liquor, chicken and pannia.[5]

Self-reliance and Outside Leadership

The change in people's confidence, their ability to intimidate the

[4] This was corroborated by something a villager from Attha overheard when he had gone to the Sondwa police station on some business. (I heard this a week later when I passed through Attha.) The police officer was describing his ordeal, how one should never venture into the hills along the river: 'The people there might have murdered us, and no one would have known. All that we got to eat that day was a dish of kulthyan ghugri. That woman, Amita, she is used to eating with a spoon; how she is living in the jungle, I don't know.'

[5] The court case against the Alirajpur administration had a satisfactory, if somewhat tepid, ending from the point of view of the Sangath. The High Court cautioned the Deputy Superintendent of Police and asked him to be more careful in the future — an order that would hopefully make the local administration more wary of tangling with the Sangath.

police who had for generations intimidated *them*, is born from their experience of successful political action through the Sangath. This change in consciousness owes a great deal to the activists who work amongst them. The presence of the activists has altered people's perceptions of the risks which made collective action seem hazardous, making the odds look less daunting. Besides reducing risk, the activists, with their access to outside resources, have expanded the choices available to villagers. Their struggle is reported in the press; they can get help from other unions and organizations. However, people did not instantly appreciate the possibilities opened up by the activists; the expansion of their political horizons was not immediately evident.

In order to persuade people, activists must demonstrate their own credibility as individuals capable of being trusted and respected. Activists are not automatically accorded authority; they have to prove themselves. Much of this proof lies in a charismatic mix of assertiveness, doing favours, and friendship. Activists display their commitment in the zest with which they work: living in people's homes, speaking their language, entering wholeheartedly into their concerns. When someone needs to go to the hospital, or has to get a document out from the patvari, they can look to the activist for ungrudging help, without any cost. Unlike other bazaarias, they put on no airs but are friendly. But they are not always easygoing; activists get their way by getting angry, by nagging and cajoling. Authority accrues to those who behave authoritatively. Such leadership has succeeded in catalysing effective collective resistance.

However, leadership is not a role with which activists are entirely comfortable. The question of dependency dogs some of them who see the development of a self-reliant adivasi-run Sangath as their goal. Ideally, they would like to move towards a situation where people have acquired *their* skills and they are rendered redundant. To a certain extent, such a transfer has happened with three of the full-time adivasi activists who have emerged as leaders in their own right. These people can handle the press in Alirajpur and Indore, lead delegations to the district administration, represent the Sangath at meetings with other non-governmental organizations, and they are widely respected in Alirajpur and beyond. However, adivasi activists are at a disadvantage when it comes to dealing with an English-speaking world — soliciting funds from organizations in Delhi, or engaging in litigation. Unfortunately, the Sangath continues to be dependent on the goodwill of supporters among the English-speaking intelligentsia, and therefore on 'outside' activists who have links with them. Apart from the adivasi activists, there is still a tremendous gap between the abilities of the activists and the villagers when it comes to dealing with the administration. There are several tasks — those that involve writing applications, knowing administrative or judicial procedures, for instance — which tend to remain in the hands of the activists. Resources outside the community — such as the press or help from other unions — are only available through the activists. Indeed, even in the case of opposing nevad on their own turf, villagers look to the activists for advice, relying on their 'better informed' judgement about the possible response of the state, and supplementing it with information gathered at meetings with other Sangath villages. However, it seems that the activists' concern about fostering self-reliance is not necessarily shared by villagers who tend to see their presence as a free service. People value the work done by the activists, but feel that they cannot take over, indeed, *need* not take over their role as leaders and initiators. They are quite content with the present division of labour. People have often expressed their enthusiasm for starting a co-operative marketing society; activists' response to this scheme has been lukewarm for they see themselves as saddled with the considerable work of book-keeping and management, more tasks that they could not delegate to adivasis.

As part of an effort to create a more self-reliant membership where people are confident of their abilities and can work independently, activists hold periodic training camps where, through discussions, role-playing exercises and games, people learn different skills and, together, work towards improving their political understanding. From a recent training camp emerged a decision to create a formal Sangath structure which decentralized responsibilities into the hands of village committees elected for different tasks. It was decided that the activists would concentrate on expanding the Sangath into new areas, while villagers would take over the running of the union in the member-villages. Paralleling the structure of the state, committees were formed in every village to look after different matters such as education, land improvement through tree plantation and composting, liquor control, feud settlement, and so on. The work of the village-level committees would be co-ordinated by a cabinet of seventeen people who would supervise Sangath-wide matters.

Activists feel that if such a formal organization takes root, the Sangath will become self-sustaining, the initiative for action will come from the people themselves. However, this hope seemed to wilt in the early days of the effort. At the first village meeting in Anjanvara, people quickly nominated members for the various committees. Some of them went to attend meetings in other villages and came back enthused. But it stopped there; the decisions collectively made were left unimplemented. When one of the activists travelled through to reiterate that the committees must work, everyone nodded their head in complete agreement. I asked a little tartly, 'You keep saying "yes, yes", why don't you *do* things then?' One man replied with great conviction, 'You know what we adivasis are like, gande through and through.'[6] I was left wondering whether that was a display of polite circumvention, passive resistance or sheer bloodymindedness.

Engendering Resistance or Providing Services?

The village committees were formed to take on many of the welfare functions of the Sangath. On starting the work of political mobilization in this region, activists were constantly called upon to fulfil the role of the missing welfare state. Given people's acute deprivation in terms of basic health and education, the Sangath could not leave these needs unaddressed. This has thrown up a dilemma for the activists: their resources are severely limited but people want them to provide services. However, the activists' efforts at service-delivery have been beset with contradictions.

One aspect of the Sangath's strategy consists of pressuring the state to live up to its public mandate. Giving in to the demands of activists, teams of government doctors have visited remote villages where they would never have gone otherwise, to vaccinate children and to treat sick people. On being asked, they have supplied medicines in bulk to activists. However, such co-operation is intermittent and idiosyncratic, dependent on the personal

---

[6] As explained before, gande can only be roughly translated; it means delinquent, mischievous, untrustworthy.

friendship between particular activists and government officials; it is not a systemic change. Overall, the state establishment remains inaccessible to adivasis because they are not consequential clients.

The omissions in the case of education are even more glaring. Most villages do not have schools; the few that do are staffed by teachers who rarely bother to attend. If, by some serendipity, a school is blessed with a conscientious teacher, there are few textbooks or other aids to learning; children learn by rote, understanding little of what they recite, most of which is irrelevant to their lives anyway.[7] The adivasis who make it through the system become estranged from their people; their qualification gives them access to government jobs and they are usually assimilated into the ranks of the exploiters. This dismal situation demanded drastic action; the Sangath rejected the government school system and instituted its own educational programme, based on teaching people to read and write in Bhilali, with politically nuanced lessons about the forest, the haat and life in the village. While it is too early to evaluate the success of this model in creating a politically educated, literate membership, its achievements are already appreciated, and it is heartily supported by the villagers.

While it has been relatively easy to replace the state by starting a small-scale alternative educational system, other projects that aim towards self-reliance have been seriously impeded by the tiny resource base of the Sangath and the hostility of the government. After a great deal of discussion, the Sangath decided to field its candidates in panchayat elections. Some people had opposed this move, arguing that the Sangath's politics would not emerge unscathed once it entered the state structure. But the majority felt that their union could not miss a chance to participate in mainstream electoral politics; by limiting their participation to the

[7] This description may sound unsubtle, but the horror of it is hard to describe. In the market town of Chhaktala, I met eight boys from Attha cramming in preparation for their eighth class examinations. Attha is one of the fortunate villages where a master teaches regularly; these boys were the few who had not dropped out of school. I was stunned to find that they were *memorizing* the answers to various mathematical problems! They were not taught that arithmetic problems had to be solved. One student told us that of the thirty students who took the tenth grade exam the previous year, only one passed. Little wonder that only 4.6 per cent of Alirajpur's population is literate (a figure that, not coincidentally, exactly tallies with the tehsil's bazaaria population).

local level, they would be able to retain control over their elected representatives. By running the panchayat, they would have access to government funds that they could spend on projects that better reflected their priorities.

The trepidations of the minority proved to be well-founded. While the Sangath won overwhelmingly in the panchayat elections, the projects that it initiated became mired in a bureaucratic morass. The activists designed innovative school and panchayat buildings that replaced the standard concrete blocks of public works architecture with low-cost local materials. These and the other projects of contour-bunding were carefully planned to have labour to capital ratios that far exceeded those of standard government projects, thus fulfilling the state-mandated objective that local projects employ as many people as possible. However, the best features of the project contributed to its failure. The administration locked the Sangath in a maze of paperwork; since the activists refused to bribe their way through, they were confounded at every turn, their accounts rejected on technicalities, their payments to villagers delayed. Today, the panchayat projects, memorials to the dead hope of participation in local government, are millstones around the Sangath's neck. The activists can only derive solace from the knowledge that their efforts at least thwarted other, less incorruptible people from embezzling local funds.

The frustrations of engaging in panchayat work bring out the dilemma faced by the Sangath. Their politics of resistance, which has been successful *against* the state, does not seem to accommodate working *with* the government. While the mobilization to protect nevad necessarily rejects the authority of the state, involvement in local government tries to subvert the state from within, but fails because of the greater inertia of the state structure. At present, the Sangath is too small to overturn the entrenched animosity of the administration towards the people. Whereas most members would like to see the Sangath take on more welfare functions, such as running the marketing co-operative mentioned earlier, the activists feel that their energies are drained when they try to infiltrate the state. The debate between agitational and 'developmental' activities continues.

However, even though some activists feel that the opportunity cost of being 'constructive' is too high, and that they are better served if they conserve their strength for organizing and mobilizing people to make demands on the state, the Sangath cannot limit itself to fighting off the Forest Department. Even if people's nevad holdings are temporarily stabilized, it is not enough. The problem of guaranteeing them security of tenure remains unresolved. Without that minimal security, without the resources to treat the soil or plant trees, the Sangath's work is only half-done, the saving of nevad a hollow victory. Under the circumstances, the activists keenly feel the need to undertake projects that will make production in the hills ecologically sustainable. To this end, they have tried to initiate forest conservation programmes through village committees. It was agreed that people would not extend their nevad holdings; in addition, parts of the forest have been cordoned off for regeneration. Villagers no longer allow people from other villages to use their forest. Yet, these efforts, commendable as they are, remain limited by the modest resources of the Sangath; much that needs to be done is simply beyond their capacity. To increase their strength, the activists have started expanding the area of the Sangath and co-ordinating their work with other groups in the region.

The activists realize that their work has only succeeded in staving off the state temporarily, and that while most people are content with this, much more has to be accomplished. Even these gains will be short-lived if the deterioration of the soil and the forest continues undiminished; the people of the hills are already migrating to the plains in search of work. However, the Sangath has engendered a change in political consciousness, and has provided an experience of organization, that will stand in good stead wherever adivasis stay or go, whether in the hills or the plains. In reviving the stilled stream of insurgent consciousness, the Sangath has effectively reiterated that power is not vested in an omnipotent, inaccessible state structure, but is reproduced through people's everyday actions of conformity or resistance. 'The more people refuse inducements, withstand intimidations, question assumptions, demand new kinds of information, repudiate values or norms of behaviour, the more powers of control are driven back upon physical coercion; and any authority which can only reproduce the governing relationships of society by a constant exercise of force is precarious' (Marris 1987: 116). And of course, quite apart from the changes wrought by the Sangath in the future shape of the adivasi-state relationship, it is immensely satisfying in itself to know that, for a brief historical moment, the adivasis of the hills came together and stood up against the state.

In this chapter we saw how the Sangath tries to engage people's attention away from the divisive politics of the feud to create an imagined community of adivasis. This transformation of consciousness has resulted in a distinct adivasi identity which propels Sangath politics. The Sangath had focused its analytical gaze on the immediate exploitative relationship between the state and adivasis. Essentially, this responded to the injustice of state appropriation but, while exercised over issues of social distribution, did not question the overall pattern of change towards a resource-intensive model of production. The knowledge that twenty-six villages in the Sangath lay in the part of the Narmada valley to be submerged by the reservoir of Sardar Sarovar dam in neighbouring Gujarat added a new dimension to the Sangath's analysis of state exploitation. At first, other than raising the problem of rehabilitation, the construction of the dam seemed to have little bearing on Sangath agenda. However, as a part of the Andolan, the Sangath activists came to relate the dam to an ecologically unsustainable and inherently inegalitarian model of 'development', which could only be opposed in its entirety. The issues framed by the resistance against the dam are discussed in the next chapter.

9

The Politics of the Andolan

Nature has always been a contested terrain for the hill adivasis. The previous chapters have discussed the hill adivasis' changing relationship with their natural environment, as mediated by the state. We have seen how conflict over different uses of nature has been expressed through collective action. While political action is not exclusively the struggle over natural resources, and people engage in feuds in defence of honour, mobilization through the Sangath has directed energy away from feuds to the conflict over natural resources. This chapter will discuss another level of political action — resistance against displacement by the reservoir of Sardar Sarovar Project (SSP).

At first it may seem as if the imminent catastrophe of displacement came as a bolt from the blue, the proposed submergence violently shaking and recasting the established configuration of relations governing people's lives. But, of course, God does not play dice. The Sardar Sarovar Project was not *deus ex machina*, but consistent with the state's appropriation of natural resources. Just as the forest has been alienated by the state, so too has the river. The SSP is one of India's many large multi-purpose river valley projects that seeks to abrogate the riparian rights of one section of the population in order to provide water or electricity to other people, mainly elites. At issue in the case of the SSP is not only the right to water, but also the larger question of displacement of communities from their ancestral lands and from their established ways of life. While the conflict in the Narmada valley has centred around displacement, there has been a simultaneous movement which critically examines the overall implications of the SSP in terms of its economic viability, financial implications, the distribution of its benefits, and its environmental impact. Together with the issue of the rights of the people in the valley, these questions strike at the very heart of the model of development that SSP symbolizes, challenging the assumptions upon which the model is based.

The opposition to the SSP has been conducted through the organized efforts of the Narmada Bachao Andolan (Movement to Save the Narmada). The struggle is being carried on at many levels, on a vast canvas stretching from the Narmada valley to the NGOs in Washington D.C. and Tokyo. I do not expect that I can fully explore all of the Andolan's enormous achievements; the magnitude of the issues raised, the range of strategies employed, and the global implications of the movement are simply too large to be adequately addressed here. I shall begin this chapter with a brief recapitulation of the history of the Project and its stated objectives, followed by a description of the people threatened with displacement. This will be followed by a summary of the Andolan's collective action, and the response of the state to its action. In keeping with my approach so far, in this chapter too, I shall maintain my focus on the village of Anjanvara, describing how its people fight against their displacement through the Andolan. Since the population mobilized by the Andolan' includes other contiguous parts of the valley, I shall describe how the people of Anjanvara have united with their neighbours in the plains of Nimar who are also threatened by displacement.

However, while the marrying of the struggles of hill adivasis and Nimari landowners has been mutually beneficial, it has also given rise to contradictions which cloud the Andolan's critique of development. I shall conclude the chapter with a discussion of the pragmatic choices that the Andolan has been compelled to make in order to fight against the dam, and shall explore whether these choices create or compromise the establishment of an alternative political culture which is the goal of the critique of development.

The History of the Sardar Sarovar Project

Although the SSP has been in the public gaze for less than fifteen years, the antecedents of the Project can be traced as far back as 1946, when the idea of harnessing the waters of the Narmada was first proposed. While the initial studies were conducted soon after, the provinces (which later became the present-day states of Gujarat, Maharashtra and Madhya Pradesh) could not agree about issues such as the sharing of the water. The Narmada Water Disputes Tribunal was set up in 1969 to resolve the conflict between the states; it submitted its report in 1978. Final planning, financial allocation and work on the Project properly started after that (Kalpavriksh 1988: 1–2).

The Sardar Sarovar Project is one part of the Narmada Valley Project (NVP), the single largest river valley project in India. The entire NVP envisages the construction of thirty major dams, of which ten will be on the Narmada and the rest on its tributaries. Of the latter, those on the Tawa, Barna, Sukta and Bargi have already been built. In addition, the NVP includes the construction of 135 medium and 3000 minor dams. The Sardar Sarovar Project is the second largest dam of the NVP in terms of area submerged and population displaced; its construction started in 1961 under Prime Minister Nehru, but gathered speed only after 1985 when the World Bank agreed to fund a part of it.[1]

[1] The World Bank entered into credit and loan agreements for $450 million in 1985, which at the time represented eighteen per cent of the cost of the dam and power project and about thirty per cent of the water delivery project. However, cost escalation and devaluation of the rupee have made it difficult to estimate the total cost of the Project. In 1985 the World Bank estimated that the project cost Rs 13,640 crores. On the basis of figures included in the Gujarat

The Sardar Sarovar dam is intended to harness the Narmada for irrigation, drinking water and power generation. The 455-foot high dam is being built at Navagam in Gujarat (see Figure 1), from where a network of canals will branch out to irrigate 1.8 million hectares of this drought-prone state and supply drinking water to over forty million people. In addition, the dam will be used to generate 1450 megawatts of energy. However, these claims have been closely examined and appear to be implausible (see Ram 1993).

The damming of the Narmada will create a vast puddle, sub-merging approximately 37,000 hectares of land. In all, the government estimates that about 152,000 people, or 27,000 families, will be directly affected by displacement due to the reservoir (Morse and Berger 1992: 62). However, those affected include not only the people in the submergence area, but also those displaced by the construction of infrastructure and the canal, by compensatory afforestation, by secondary displacement and more. When these people are added to those living in the submergence area, the numbers swell to more than a million (Ram 1993: 1). Incredibly enough, since most of these people have not been classified as 'project-affected persons' or 'oustees', they are not entitled to any compensation. The compensation package for those in the submergence area who *are* classified as 'project-affected' varies from state to state; it also appears to be extremely unlikely that it will ever be delivered. Several recent studies have shown that resettlement has considerably worsened the lives of the few families that have shifted so far; there is every indication that satisfactory rehabilitation is impossible even for those offi-cially classified as 'project-affected people' (TISS 1993; Bhatia 1993).

Of the 245 villages to be submerged by the reservoir of the dam, 19 lie in Gujarat, 33 in Maharashtra and 193 in Madhya Pradesh. While the 37,000 people in the villages of Gujarat and Maharashtra are almost all hill adivasis, the submergence area in Madhya Pradesh consists of two distinct zones. The uppermost reaches of the reservoir will flood land in the plains of Nimar, part of Dhar and Khargone districts of Madhya Pradesh. Here

---

state budget, the total cost in 1992 terms should be around 20,470 crore rupees (Ram 1993: 34).

the Narmada flows through a fertile valley, settled by a mixed population dominated by Hindus of the Patidar caste. Two-thirds of the displacement due to the dam will occur in Nimar. While roughly a third of these 100,000 people are adivasis, classified as Scheduled Tribes, they have largely been assimilated into the dominant Hindu caste structure, and do not share the distinctive cultural identity of their cousins in the hills. Downstream of Nimar, the river cuts through the Vindhya and Satpura mountains, forming the boundary between Maharashtra and Jhabua district of Madhya Pradesh (of which Alirajpur is a part). The river here is bounded by steep hills, fissured by the paths of tributaries into escarpments and shelves where, on the north bank, live the hill adivasis of Alirajpur. Approximately 15,500 of them, including the population of Anjanvara, will be affected by submergence. From here, the river flows through to the dam site in Gujarat and then to the sea.

Although the government plans to drastically alter the lives of so many people with the Sardar Sarovar Project, it has not bothered to consult or even inform them about their fate. In Anjanvara, the first information about the dam came from the Central Water Commission surveyors who came to place stone markers to indicate the reservoir level! According to the Schedule of Submergence released in March 1991, Anjanvara's fields and houses will be temporarily submerged in 1994; in 1995 they will be permanently under water (NCA 1991: 14). No one from the government came to tell the people of this deadline; they heard about it in a Sangath meeting the following month. Meanwhile, Anjanvara lies in limbo, with developmental work suspended. A handpump for the village was sanctioned several years ago, but never installed because the village lies in the submergence zone of the dam. So the last few years and the present have been held captive to an uncertain future.

Despite the large numbers of people affected, despite the enormity of the change in their lives, there is *no* government-sponsored system of information that respects people's right to know. People glean scarce facts from informal conversations with the patvari (revenue official) or similar officials when they meet on other business. In 1986, an independent survey discovered that, contrary to the claims of the government, no one in Alirajpur had been issued Land Acquisition notices (MARG 1986: 5). As far as I could ascertain, the people of Anjanvara have *still* not been formally informed that their land is being acquired by the state. Villagers have also not been apprised of their rights under the terms of the Narmada Water Disputes Tribunal award. Clearly, the planning and implementation process of the Project does not include the participation of the affected population and is a callous violation of their right to information, cultural autonomy and choice.

The Narmada Bachao Andolan: A Summary of Action

The first stirrings of protest against the SSP started in 1978 in Nimar soon after the Narmada Water Disputes Tribunal gave its award. This was the year when the Janata Party was in power at the Centre and in Madhya Pradesh. Arjun Singh, the leading Congress (I) politician in Madhya Pradesh, mobilized people in Nimar around the issue of displacement in what came to be known as the Nimar Bachao Andolan (Movement to Save Nimar). Although this campaign was chiefly supported by merchants and farmers in Nimar, and worked within the established structures of party politics, the attendance at its rallies is said to have been much larger than anything seen today. However, it transpired that Arjun Singh was only interested in milking the issue to further his own political fortunes; he won the state elections in 1979 on a platform that pledged support for the movement, and then promptly ditched it. After being betrayed by Arjun Singh and the Congress (I) apparatus, the movement collapsed.

The second attempt at organizing the opposition to the Project occurred around 1985,[2] when Medha Patkar, a social scientist from

[2] This was also the time when environmentalists outside the valley expressed their disquiet about the NVP in the press (see Kalpavriksh and Hindu College Nature Club 1985). The commitment of the World Bank and bilateral aid agencies to fund the SSP also occurred then, setting off an international wave of protest at a time when there was mounting evidence about the dark side of large dams (see Goldsmith and Hildyard 1984). The anti-dam cause was endorsed by the Indian Ministry (then Department) of Environment and Forests which said that the impact of the two largest dams, SSP and Narmada Sagar, had not been adequately studied, and refused to give the Projects environmental clearance. However, the Ministry was overruled by Prime Minister Rajiv Gandhi who agreed to let the required studies be conducted *pari passu* with the construction of the Project. Thus, the Project was allowed to proceed even though its impact was imperfectly known.

the Tata Institute for Social Studies in Bombay, who had become involved in community mobilization, started working in the SSP submergence zone villages in Maharashtra. The anti-dam mobilization in the affected villages of Alirajpur was left to the Sangath which had an established base among them. Like the people of Alirajpur with their experience of the Sangath, those of Maharashtra were familiar with the Shramik Sanghatana (Workers' Union), and were consequently somewhat easier to mobilize. In 1987, Patkar came to Nimar to build opposition to the SSP.

While the NVP is a massive scheme which has to be resisted as a whole, it has not been possible to locally oppose all the dams. The Andolan has concentrated its efforts on collective action against the two largest dams of the NVP — the SSP and its upstream companion, Narmada Sagar. In the last three years, as the future of Narmada Sagar has begun to look more and more uncertain, and work on Sardar Sarovar has gathered steam, the opposition has mainly been focused on the latter. Initially, the Andolan did not challenge the overall validity of the SSP, but started with the intention of organizing people to agitate for adequate rehabilitation.[3] However, when it became apparent that it was simply impossible for the state to properly resettle all the Project-affected people, and moreover, that the Project was questionable on other grounds too, the Andolan changed its position to a total rejection of the Project, voiced in the slogan *Koi nahin hatega! Baandh nahin banega!* (No one will move! The dam will not be built!)

[3] Indeed, the rehabilitation policy of SSP, advertised by the government as the most liberal ever, is the outcome of concerted activism in the early years by voluntary agencies such as the ARCH-Vahini and the Rajpipla Social Service Society in Gujarat. These organizations, together with the Centre for Social Studies in Surat, helped formulate principles of rehabilitation that were, for the first time, accepted by the government (CSE 1985: 106–9). In a major breakthrough in state policy, the Gujarat government agreed to give land in the command area of the Project in compensation for land acquired by the state, and also promised to give land to the landless. Later, the public pressure created by the Narmada Bachao Andolan compelled the government to announce even more generous terms for the 'oustees'. However, this rehabilitation package, fairly progressive in comparison with what went before, still remains inadequate for compensating people for all their losses. Worse still, the provisions of even this policy have not materialized for most of the people in the submergence zone.

. The campaign against the SSP has been conducted on all fronts. Popular mobilization in the valley has been supplemented by co-operation with three broad categories of NGOs and mass movements: within India, city-based NGOs have participated in the Andolan by disseminating information about the movement through press briefings, newsletters and films, by lobbying, collecting funds for the valley, and by organizing events to express solidarity with the struggle in the valley. Rural-based mass organizations have sent their members to participate in Andolan campaigns, a gesture of co-operation that is reciprocated by the Andolan. The national canvas of the Andolan is encapsulated in the structure of its co-ordinating committee which consists of representatives from all over the country, albeit with a central and western Indian bias. The Andolan also receives the help of western NGOs which are engaged in pressuring the international financial community to withdraw its support from the Project. In 1989, three US-based environmental NGOs — the Environmental Defense Fund, Environmental Policy Institute and National Wildlife Federation — urged the United States' Congress to compel the World Bank to stop funding the SSP. Their lobbying succeeded; in an unprecedented decision, the Bank appointed an independent team to review the environmental and displacement impact of the Project. In June 1992, the Independent Review submitted a highly detailed report which was consistently and strongly critical of the Project, as well as of the Bank for funding the SSP in violation of its own guidelines (see Morse and Berger 1992). Continued pressure by international NGOs made the World Bank's involvement in the Project increasingly embarrassing and, in March 1993, the Bank stopped funding the SSP. Similarly, the persistent efforts of Friends of the Earth in Japan convinced the Japanese government, usually impervious to environmental concerns about the projects that it finances abroad, to suspend aid to the SSP. International recognition for the Andolan's cause has come in the form of the Right Livelihood Award, given in Sweden in December 1991, and the Goldman Award, given in the United States in May 1992.

The Andolan has also initiated litigation in state courts challenging improper land acquisition and forcible eviction, state repression, and denial of the constitutionally-guaranteed right to life. The arguments against the SSP have been buttressed by independent research which has questioned the validity of its benefit-cost analysis (see Paranjpye 1990; Ram 1993) and which has proposed less devastating alternatives to the Project. The Andolan has gained the support of celebrities ranging from film stars to Supreme Court justices, including many prominent social workers. The most renowned among them is Baba Amte, famous for his work with leprosy patients. As far back as in 1985, when dams had not become synonymous with controversy, and even the far-sighted State of India's Environment declared that 'as things are today, people are not likely to oppose dams en masse' (CSE 1985: 102), Baba Amte led a mass movement against the construction of Bhopalpatnam and Inchampalli dams in Bastar, Madhya Pradesh, because they implied the 'ethnocide' of the adivasis of that area. In July 1988, Baba Amte held a historic meeting at his ashram in Anandwan where environmental activists from all over India came together to forge a common policy on dams. In March 1990, Baba Amte left his ashram in Maharashtra and moved to live in the Narmada valley, declaring that he would die there fighting the dam. Baba Amte's newly built ashram in the valley is testimony to hope in life; he has planted banyan saplings — slow-growing trees that take years to reach their majestic size — an assertion of the belief that the valley will not drown. These diverse tactics, aimed at several different audiences, have been employed as part of a co-ordinated strategy to fight the dam at all levels — the international financial institutions, the national and the state governments, and that mysterious but powerful force, public opinion.

Protest in the Narmada Valley

The gathering forces of opposition against the Sardar Sarovar Project have their centre in the valley, mainly in Maharashtra and Madhya Pradesh, among the people threatened with displacement. Since 1988, the inhabitants of the submergence area have been demonstrating their determined refusal to move from their land. The last five years have been marked by continuous protest; no survey work related to the Project has been allowed to proceed. The submergence markers placed by surveyors were pulled out and sent to the state capital. Officials of the Project, including visiting World Bank teams, have been mobbed and sent back.

People have held demonstrations and launched relay hunger strikes outside Project authority offices, forcing them to close. The construction of bridges and guest houses has been delayed. Even though the dam is being built and the scheduled submergence looms near, people have, by and large, not moved from their homes.

The constant passive resistance to the Project has been punctuated by larger demonstrations, exhibiting the Andolan's strength in the valley. The first of these occurred in February 1989 when more than 8000 people gathered at the dam site in Gujarat to protest against the Project. Many were beaten and arrested by the police. From this beginning, the Andolan went on to organize what has come to be known as 'the coming of age of the Indian environmental movement' — the National Rally Against Destructive Development held on 28 September 1989 in Harsud, a small town in the submergence zone of Narmada Sagar Project. This historic event was attended by 25,000 to 60,000 people from all over India who

forcefully demanded an end to all projects which devastate the environment and destroy people's livelihoods, and called for the adoption of a socially just and ecologically sustainable pattern of development. Its defiant message, to politicians and planners, was that people are no longer prepared to watch in mute desperation as project after destructive project is heaped on them in the name of development and progress ... And so on September the 28th, 1989, people struggling against past and proposed displacement and environmental degradation, by massive irrigation and power projects such as Sardar Sarovar and Narmada Sagar, Bhopalpatnam-Inchampalli, and Koel Karo, defence projects such as Baliapal, nuclear power plants such as Kaiga, came together in an unprecedented show of strength (NBA Delhi 1990: 4).

The highlight amidst the fiery speeches, torchlight processions and tribal dances at Harsud was the *sankalp* (resolve) — a pledge opposing the displacement of poor communities and the plundering of their natural resources by destructive development which served the interests of elite consumers and foreign capital. The rallying cry at Harsud, which echoed through all subsequent events, was *Vikas chahiye, vinash nahin!* (We want development; not destruction!), a cry that reverberated throughout the world. The assembly of progressive groups from all over the country at

Harsud also led to the formation of a federation called Jan Vikas Andolan (Movement for People's Development), a structure to facilitate the exchange of ideas, mutual co-operation and co-ordinated action.

After the euphoria of Harsud, the next major episode of symbolic protest occurred on 31 December 1989 in the village of Hapeshwar where thousands of hill adivasis gathered in a stirring ceremony and held aloft their bows and arrows and, with the waters of the Narmada cupped in their palms, pledged to fight the dam until their dying day. A little more than two months later, on 6 March 1990, 10,000 people from the valley assembled on the bridge over the Narmada at Khalghat, blocking traffic on the Bombay–Agra national highway. Their day-long takeover of the bridge crippled the movement of trucks and buses on this important route. The rally was withdrawn when the Chief Minister of Madhya Pradesh promised to review the SSP. However, that assurance was promptly and conveniently forgotten by the government.

With construction of the dam wall continuing unimpeded, the Andolan has had to race against time, organizing ever-larger, increasingly dramatic protests. These have impressively demonstrated the Andolan's logistical capabilities in mobilizing resources at all levels.[4] In December 1990, the Andolan set off from Nimar on the Jan Vikas Sangharsh Yatra (March of the Struggle for People's Development) to peacefully stop work at the dam site. The Andolan was demanding a comprehensive review of the Project, pending which all irreversible work on the river bed and clearfelling of forests should be suspended. In solidarity, a group of four hundred representatives from tribal mass organizations met the President and the Prime Minister in Delhi and asked them to intercede in the matter. However, the two kilometre-long procession of people, singing and chanting slogans, marched for only six days before it was stopped at the interstate border between Madhya Pradesh and Gujarat by a massive police blockade.[5] Unable to proceed towards the dam site, the Andolan camped at the border in non-violent satyagraha. About two thousand people

[4] For stirring eyewitness accounts of the organization of the Sangharsh Yatra and the Manibeli satyagraha, see NBA Delhi (1991, 1992).

[5] For an account of the state response to the Sangharsh Yatra, see Baviskar (1991).

stayed at the Gujarat border for more than a month, braving the winter cold by sleeping in an encampment in the open fields.

During the Yatra, in order to pressure the government even further, seven people including the leader of the Andolan, Medha Patkar, started an indefinite hunger strike. As the condition of the fasting protesters started to deteriorate, the police swooped down at night to take them into custody but were deterred by the spirited opposition of the people. Instead of negotiating with the Andolan, the Gujarat government launched a vicious press campaign to malign the movement, backing it up with arrests of people trying to cross the border. After twenty-two days of fasting, when the government refused to talk, the Andolan retreated from the border, vowing to take the struggle back into the villages. A programme of non-cooperation with the state was declared under the slogan *Hamaare gaon mein hamaara raj* (Self-rule in our villages).

The rising dam wall has started impounding water. While the water flows a little sluggishly now, it is still below the level that will permanently submerge villages. However, the river has always been prone to flashfloods during the monsoons, when the water suddenly rages forty to fifty feet higher than usual. Now that the dam straddles the path of the water, houses built above the normal flood line stand in danger of being swept away during the months of rain. Manibeli, the Maharashtra village closest to the dam site, faced such a threat during the monsoons of 1991. The Andolan shifted the focus of its struggle to Manibeli, asserting that since people had repeatedly expressed their resolve to die rather than leave their land, they would court death in Manibeli. A *samarpit dal* (band of the dedicated), also known as 'save or drown squad', was stationed in Manibeli from 14 July 1991 to 26 August 1991, where it joined local villagers in satyagraha. The determination of the protesters impressed even the hostile Gujarat press; their resolution forced a nervous government to desperate action, arresting people and imposing prohibitory orders against public assembly. Despite police repression, the satyagraha continued until the waters receded. Although the high drama of Manibeli in the monsoons ended without tragedy, it was dangerous brinkmanship; the state seems to be determined to prevent such a showdown this year. The summer of 1992 has already been marked by several police attempts to forcibly evict the people who continue to live in Manibeli; the events in this year's monsoons may be a watershed for the Andolan and the fate of the Project.[6]

State Repression Against the Andolan

During the last four years, collective action in the valley has continued despite constant state repression. Although the Andolan has expressly avoided direct confrontation and militant protest, even peaceful demonstrators have had to deal with police harassment, arrests and violence (Sangvi and Agrawal 1991). In January 1989 the government sought to stifle the expression of dissent against the Project by making the entire dam site a prohibited area under the purview of the Official Secrets Act. Defiant people who tried to enter the area were arrested. When protests at the dam site were banned, people started agitating in the submergence area and were met with frequent police beatings and arrests, the constant, wearying irritation of threats and false cases, and the obstruction of everyday life. Women, who are strategically placed at the forefront of most demonstrations, are specially subjected to brutal assaults; their clothes are ripped off in public, they are dragged along by their hair — in one incident, a pregnant woman was repeatedly hit on her stomach with a rifle butt (PUCL 1990).

Since 1991 police violence has escalated in the valley as the state has chosen to start terrorizing people into leaving their villages. Survey teams have been accompanied by the police who have often swung into action to prevent protest, their actions designed to intimidate people. Besides suppressing peaceful demonstrations and non-violent acts of civil disobedience, the government has tried to break down the resistance in the villages by bribing village leaders and by threatening them with the power of the state. One such incident occurred on 30 October 1992 when the Collector of Jhabua, an officer of the Indian Administrative Service (IAS) and the head of the district administration, held a 'problem resolution camp' in Kakrana, a large village in the

---

[6] Since the time of writing, the remaining villagers of Manibeli and Vadgam were forcibly evicted. In the summer of 1993, the Andolan's strategy of indefinite fasts and the threat of *jal samarpan* (drowning) received massive press coverage, resulting in the creation of a tidal wave of public support for the movement. This forced the Indian government to give in to the demand for a comprehensive review of the Project.

submergence zone. Like Anjanvara, Kakrana is one of the seven-teen villages of Madhya Pradesh which is scheduled for submer-gence in the first phase of the Project, and has been the scene of a number of confrontations between the authorities and villagers opposing the dam. Instead of discussing local problems in detail, the Collector told assembled villagers that the combined might of the state — lathi, bullet and pen — would be used to persuade people to move from their land. Later, as the people crowded around the Collector's car, demanding to be told where there was land for resettlement in Madhya Pradesh, the officer slapped a Sangath activist several times and ordered the police to arrest the leaders of the group. Subsequently, five arrested men, all active in the Sangath and known to the police, were handcuffed and taken to the Alirajpur police station. Around midnight, three police officers and the Sub-Divisional Magistrate (SDM) of Alirajpur entered the room where the villagers were being held and shut all the doors and windows. Then the SDM, an IAS officer, told the policemen, 'You people don't know how to beat properly. I will show you how it's done.' One of the activists described what happened next:

The SDM unlocked my handcuffs and then, with the aid of two police officers, grabbed me by my hair and threw me to the ground. As I lay on my stomach, he grasped my legs and arched them back towards my head. Then, while my body was twisted in this painful arc, the SDM sat down on my legs. I screamed with agony. Then the Sub-Divisional Officer (Police) intervened and told the SDM, 'Sahib, don't hit him like this.' After this the SDM turned me over and, with the help of two policemen, tied my raised feet to a bench. I was hit on the soles of my feet with a lathi about twenty times. The SDM told me, 'So you fancy yourself as a leader? Carry on with your *netagiri* (acting like a leader) and you will die just as Shankar Guha Niyogi did.[7] We will dump your body in a lake. Who do you think you are?' I fainted with pain. When I came around, the SDM ordered me to stand up and march in place. As I did what he said, he kept hitting me on my legs and back with a lathi, and calling out, 'Faster! Faster!' One by one, the other four were also beaten in this way.

[7] Shankar Guha Niyogi was the leader of the Chhattisgarh Mines Shramik Sangathan, a large trade union in eastern Madhya Pradesh, who was assassinated by hired killers in September 1991. The Central Bureau of Investigation found that local industrialists were responsible for his death.

After a couple of hours, the five men and a woman activist, who was also arrested but had been separately detained, were taken to another police station in Chandpur in a van with the policemen beating and kicking the villagers all the way. In Chandpur, the men were forced to take off everything except their undergarments and spend the night in a cold cell without any food. The next evening, the arrested people were taken for a medical examination in Alirajpur. Though they showed the marks and bruises on their bodies to the doctor on duty, he refused to record their injuries. When the group was presented before a judicial magistrate, they complained to him about their torture in police custody and showed him their injuries. The magistrate ordered another medical examination but, again, the doctor on duty refused to record their injuries. The group was released on bail four days later on 3 November 1992.

Although the Sangath activists were arrested on 30 October all the charges against them were related to an incident that allegedly took place in Kakrana on 15 October. One of the five activists, who was arrested and sadistically beaten by the SDM in the Alirajpur police station, was far away in south India on 15 October, attending a conference of tribal activists in Kerala. The activist has train tickets, photographs and letters from about twenty people from all over India vouching for his presence in Kerala. Another person in the group, a student, was attending college in Alirajpur on that day and could not possibly have been in Kakrana. His college attendance register and his hostel mess register both attest to this fact. Thus people active against the Project are arrested on trumped up charges and beaten brutally. They are harassed and their work hampered by being constantly trapped in a wearying round of police torture and humiliation, jail and court.[8] Since 1992, there has been a sharp increase in such incidents of violation of human rights in the valley; villagers who have resolved to not leave their land are subjected to a barrage of threats and

[8] The Kakrana incident and other instances of state repression against the Narmada Bachao Andolan have been graphically documented by the Narmada International Human Rights Panel, an independent consortium of 43 human rights, environmental and indigenous rights organizations in sixteen countries (see LCHR 1993). Indian civil liberties groups have also played a valuable role by constantly monitoring and publicizing human rights violations associated with the Project.

blandishments, actual violence and implied terror. With the mounting urgency of deciding the issue one way or another, the state seems resolute in suppressing opposition to the Project.

The Andolan in Anjanvara

As described before, the people of Anjanvara were not informed or consulted about the construction of the dam that threatens to destroy their present place and way of life. Through the Sangath, some men travelled to Gujarat to inspect the land that the government offered as compensation; they were not reassured by what they saw. They say that they would rather move up into the further hills than go to Gujarat. Like the other submergence area villages of Alirajpur such as Kakrana, Anjanvara has participated in the Andolan through the structure of the Sangath. That is, mobilization in Alirajpur is done by Sangath activists who work together with the Andolan activists. Villages in the Sangath which do not face displacement also join Andolan protests.

For the most part, Anjanvara's involvement in the Andolan consists of turning out in large numbers for all mass demonstrations. At Hapeshwar, they were among the many who pledged in the name of Narmada to fight the dam. During a smaller meeting in Kakrana, people from Anjanvara were among those who took a pledge in the name of juvar mata to not move from their homes. During the Sangharsh Yatra, a rotating contingent from the village stayed at the Gujarat border throughout the month-long satyagraha. A few men from Anjanvara, who take the lead in Sangath matters, also shoulder the task of mobilizing others for Andolan programmes by holding village meetings and collecting funds. They have travelled to distant places such as Delhi and Bombay to protest outside state offices, experiences that have considerably increased their self-confidence in dealing with bazaarias.

Greater familiarity in dealing with bazaarias has also made some adivasis sophisticated spokesmen for their cause, able to indulge journalists and others who demand attention-grabbing quotes and picturesque behaviour. Many Andolan events involving adivasis seem to be stage-managed with this purpose in mind — with bows and arrows silhouetted against the sky, and the frequent pledges with hands full of juvar or Narmada water. Such symbolic references resonate with bazaaria images of the adivasi, and are judiciously adapted by adivasis for that very purpose. At a meeting before the Sangharsh Yatra, an active and influential supporter of the Andolan from Delhi gave a speech in which the audience were told that, as adivasis, they were 'mother earth's children' and that the dam would tear them — the children — from their mother's breast. As a rule, adivasis never use this metaphor, preferring more prosaic descriptions of the threat to their land and livestock. However, this speech evidently made an impact because, later on, a man from Anjanvara, who is active in the Andolan and who had never before expressed himself thus, repeated this statement to a television interviewer as if it were his own. His tactic was well-chosen; another film-maker who had earlier taped an interview with this adivasi and had only heard of land being drowned, came to know of this, more colourful response. So he went back to the adivasi to rerecord the interview, this time ensuring that he had the mother-child metaphor on tape! Of course, the adivasi's willing adoption of such an image meant that it resonated to some extent with his perceptions about his objective reality, but it was also prompted by a desire to obligingly fulfil the expectations of bazaarias, a strategy that is actively promoted by the Andolan.

However, all collective action against the dam is not mediated by the presence of the press or outside Andolan activists. The people of Anjanvara have resisted the Project on their own terms and on their own grounds. On 14 May 1990, a team of surveyors and forty armed policemen on horseback came to Anjanvara (see PUCL 1990). It was the middle of the day and most men were away in the fields. On the pretext of holding a meeting, they assembled all the women in one house and then posted guards outside so that they could not leave. Later, when the women were let out, they saw the surveyors and ran to snatch away their tapes. The police forced them back with their lathis. A young girl was trampled in the stampede which followed. Several women were left beaten and bruised. However, they succeeded in preventing the survey from being conducted. The most violent confrontation between the state and those opposing the dam occurred in January 1993 in Anjanvara (see Epilogue) when the police fired several rounds of ammunition, beat up women and children, vandalized their homes and arrested around thirty activists who were beaten, tortured and humiliated in police custody.

The Andolan in Nimar

Since two-thirds of the displacement is to occur in the plains of Nimar, mobilization in this region has justifiably been a crucial component of the Andolan strategy. Even though the site of confrontation has shifted in the last year to the Maharashtra village of Manibeli closest to the dam site, the movement continues to be highly active in Nimar. As mentioned earlier, the people of the plains had participated in the Congress Party-led Nimar Bachao Andolan during the 1970s, only to be betrayed by their leader. Unlike the hill adivasis, who are only marginally interested in electoral politics, the people of the plains participate vigorously in party politics. The Bharatiya Janata Party and the Congress are the most heavily supported parties; in recent years the former has won elections at all levels. The Andolan has bypassed party politics by focusing on the Project as an issue that transcends partisan loyalties.

The widespread mobilization in Nimar is primarily the achievement of Medha Patkar who started work here in 1987. People told me that when Patkar first came to Nimar, they were wary of her, unsure if she was trustworthy or whether she would turn out to be another power-broker who would sell them out as the Congress politicians did a decade ago. Once they had verified her antecedents (it was not quite clear to me how), they accepted her. As Devrambhai, an Andolan activist from the village of Kadmal, explained, 'We had never heard of anyone who did social service.' For him, Patkar's selflessness was proof enough. So much so that today, in the space of only a few years, the man who learnt about 'social service' from Medha Patkar, wants his daughters to do the same thing when they grow up.

To the people of the valley who are dedicated to the Andolan, Medha*jiji* (sister) as she is called, is not just a leader but a little bit of a goddess, in whose power they have faith. Her charisma — part magic, part engineering — mobilizes people, transforming the equation between resources and collective action. People's allegiance is as much to her personally, as to the changed ideology or collective consciousness of the need to fight that she has brought about. As Sitarambhai, another activist from Kadmal village said:

She is a woman, yet she has so much rage within her, so much magnetism. If the people of Gujarat [those who oppose the dam] could hear

her — there is such truth in what she says that they would be compelled to agree. But they are taken up in their dishonesty. Medhajiji has given her life for the valley. But the people here — let alone fighting for others, they aren't even ready to do anything to protect their *own* selfish interests. Whereas it is the valley that is our existence and being; there is nothing beyond this. I talk about the valley night and day but it has no effect on people.

The village activists in Nimar are deeply committed to the Andolan, even though for many of them it has meant making considerable personal sacrifice. Sitarambhai's situation is not comfortable: 'My economic circumstances are such that if I do not supervise my land properly this year and the next, let alone the future, I will drown in the present.' Yet, at the same time, the idea of jal samadhi [9] is growing powerful in his mind: 'Jiji, one has to

---

[9] During the Manibeli satyagraha, a *samarpit dal* (band of the dedicated) composed of volunteers from all over the submergence zone, sat in the first village waiting for the waters to rise, joining the people of Manibeli in refusing to leave the valley. This was referred to as jal samadhi, death by water. However,

die one day or another. What can be better than that I be a martyr for the valley? It is a small sacrifice for our fight.' He has announced his decision at home; that is why his wife is stopping him, reminding him of his household responsibilities. 'That is an excuse for stopping me. I have told them what they should do when I am gone; sell this house and build a hearth in what is now the big cowshed. And sell four acres of land. Then there will be no problem in getting by.'

I said that thinking about jal samadhi scared me. Sitarambhai replied simply, 'It is better to die than to live in humiliation. We have said so often that we are prepared to drown. The moment of truth has come. We will have to go to die.' I asked, 'What if the state is not impressed by your sacrifice? What will be the effect on the people in the valley, on the Andolan?' Sitarambhai said, 'We will have to take this step without being swayed by the thought of others. If there is the spirit of adventure in anyone's heart, then they will be ready to fight. One day or another, people's atman will awake.'

Sitarambhai is deeply religious; his conversation is filled with references to Hindu epics, folk tales and parables, and he brings a religious conviction to his work with the Andolan. He has a powerful voice and sings bhajans (devotional songs) with the Kadmal congregation. Like other Hindus in the valley, he reveres the Narmada. In Nimar, on the day of the new moon, women bathe in the river and offer coconuts to her.[10] During the rains, people assemble in the village of Koteshwar where the epic Narmada Puran is recited over seven days and seven nights. The people of the plains, just like those of the hills, have strong affective ties to present communities, to the land, and to their river, though these are expressed differently, and have their bases in different environments, social conditions and religions.

Attachment to the land and the river, together with the transformation brought about by Patkar that stresses Andolan identity over party or caste affiliation, has created a cadre of dedicated

---

the term samadhi is a subtle one, implying not death, but 'a state of contemplation into which a perfected holy man is said to pass at his apparent death' (Oxford English Dictionary). Thus, death at Manibeli was not looked upon as suicide, but as martyrdom.

[10] For other examples of Hindu observances around the river, see Chapter 4.

activists in Nimar. The Andolan has also enabled the women of Nimar, traditionally jealously cloistered, to come out of their homes and take to the streets, demonstrating in front of Project authorities' offices, raising slogans, challenging the police and taunting bureaucrats and politicians. This revolutionary change has been brought about by Medha Patkar and other women activists of the Andolan. They inspire women with a vision of what they can politically accomplish outside the home; their presence reassures men that women are 'safe'. The sensitivity of the Andolan to the gendered nature of its constituency is one of its greatest achievements.

For young men, the Andolan has widened the political spectrum: involvement in the Andolan is a welcome alternative to the fundamentalist politics of the Bharatiya Janata Party or the traditional Congress. They embark on bicycle rallies, singing Andolan songs (many of them bhajans in praise of the river); some of them have been taught to use puppets and perform their own shows about the Andolan. They have started a reading room where they receive books and magazines from organizations affiliated with the Andolan. With time, some of the activists from the youth wing have matured into full-time workers, seasoned in the arguments against the Project, capable of independently managing logistics for the Andolan, who have also started charting an ideological course that is autonomous from their elders. Leaders from the youth wing, whose participation in the Andolan has exposed them to diverse mass organizations and movements across the country and abroad, are much more critical of communal politics, of caste divisions in their villages, and sensitive to issues such as gender inequality, environmental destruction and so on.

The Andolan's strength in Nimar is concentrated among the Patidars who, as upper-caste landowners, stand to lose the most from displacement. This factor gives rise to certain contradictions in the Andolan's ideology about development as socially unjust and ecologically unsustainable. The Patidars figure among India's politically powerful middle and rich peasantry (see Byres 1981). Although their lands have always been fertile, their productivity rose tremendously during the early 1970s after electrification. The fields above the river, which had earlier been watered with wells, are now abundantly irrigated with water drawn by electric pumps. Plentiful water has enabled many farmers to grow remunerative crops such as bananas, papayas, cotton and sugarcane besides the more traditional chillies and groundnuts. Not surprisingly, this class has benefitted from state subsidies in the form of cheap water, electricity, fertilizers, credit and price supports.

The highly capital-intensive nature of agriculture in Nimar enmeshes most farmers in a web of debt. Cultivators talk of the need to invest large sums of money which they borrow at an interest rate of 25 per cent per annum from the moneylenders. This money is needed to keep farming operations solvent, and for all aspects of the agricultural process from setting up irrigation pipelines to spraying pesticides. The burden of making timely payments to the moneylender, scraping together enough money to remain creditworthy, while speculating on unpredictable crop prices, makes agriculture even more hazardous than it usually is. Smaller landowners do not have the economic cushion that would allow them to sit out unattractive prices, or to grow the more profitable fruit crops which have longer gestation periods. The system is skewed in favour of larger farmers.

Listening to small Patidar farmers in Nimar, those with five to ten acres of land, one gets an impression of an overwhelming debt burden which will slowly drive the cultivator into the ground. Yet the same people also narrate a different tale — of increased assets, improved standards of living and overall optimism. The story of one farmer in Kadmal is typical: Three years ago, he and his three brothers had twelve acres of land and two pairs of oxen. Then the farm was partitioned. Today, he alone has fifteen acres of land valued at Rs 25,000 per acre and three pairs of oxen.[11] In addition, he has been able to borrow Rs 20,000 and lease in another twelve acres. Although he is extensively in debt, he is confident that if this year's crop is good, and if he gets a good price for it, his economic situation will be secure. He told me that he envisages that, three or four years into the future, his life will be comfortable. He has only one son; the land will not be partitioned further, and his son will inherit valuable, improved land.

The perception that life is improving and that future prospects look promising for those who are willing to work hard, seems to be generally held not only by the landowners, but also by

[11] Of course, since it lies in the submergence zone, this land will not fetch such a price in the current market.

labourers. They say that even as daily wage employees, they get work through the year, so their earnings have risen. They buy grain at subsidized prices, and can now afford to purchase durable consumer goods such as bicycles, radios and wristwatches — items that they point to with evident pride. In contrast to the women of the hills who still grind flour by hand every day, households in the plains have been getting flour ground cheaply from the local mill for the last twenty years. The Nimar of today has experienced a change in consumption patterns as a consequence of post-electrification and post-irrigation prosperity. People reminisce that earlier, eating wheat was a special treat. Tea was drunk only in the homes of the rich; no one had seen sugar. Now all these things are commonplace.[12]

The resources of Nimar's Patidar farmers provide the Andolan with its material base. During the Manibeli satyagraha, for example, Nimaris sent provisions for the hundreds of people camped there. They raised money by collecting donations of bags of grain and selling them. Two volunteers with motorcycles were permanently stationed to ferry people to and from Manibeli. The task of mobilizing resources is efficiently distributed among different tehsils and the villages within them. The Nimari capability and willingness to provision the campaign machinery has been crucial to the success of the Andolan, allowing it to chart a course independent of external funding.

Class Base of the Andolan in Nimar

In its fight against the Project, the Andolan has successfully invoked the issue of social justice, by persuasively arguing that the

[12] Nimar's prosperity also expresses itself in lavish hospitality, which I sometimes found embarrassingly generous. Nimaris who have motorcycles or bicycles think it *infra dig* to walk; catching the bus is just barely acceptable to them. The first time I set off from Kadmal to Badwani, a pleasant walk of nine kilometres, I was stopped and offered lifts by one tractor, one van and two motorcycles! At a village on the way, the *paan*-shop owner who is active in the Andolan asked, 'Jiji, should I take you on my bicycle?' Upon my return the next day, a concerned Bhagwanbhai of Kadmal asked, 'Jiji, are you short of money? If you are, let us know.' I assured him that I walked out of choice, but he looked dissatisfied. When I told the activists in Badwani about this incident, they warned me in mock alarm, 'Don't go putting ideas into people's heads or they will start expecting *us* to walk everywhere too!'

dam will generate benefits for an already-privileged few, while further impoverishing those who are already disadvantaged. Such a distribution will worsen existing social differences. However, while this argument is admirably apposite in the case of the hill adivasis, its applicability is somewhat uncertain for the Nimari majority in the Andolan. As discussed before, the Patidars, who are dominant landowners in Nimar, have been most active in the Andolan. Landless labourers, who constitute 40 per cent of a village such as Kadmal, and who mainly belong to Scheduled Castes and Tribes, are markedly absent from protests and from the ranks of local activists. Even though they stand to lose their livelihood and community too, they maintain a distance from the Andolan. Kadmal has a few adivasi landowning families who participate in Andolan demonstrations, but only do so when pressed by their Patidar counterparts. For villagers who are active in the movement, the key issue is the loss of their land.

The issue of land has united two disparate constituencies. Class conflict is temporarily submerged in the alliance between the adivasis of the hills and their traditional foes, the Patidars. The cause of displacement by the dam is more important than anything else; energies are concentrated on co-operation towards fighting the Project. This alliance has been remarkably successful, with no overt hostility or tension between hill adivasis and plains Patidars.[13] Hill adivasis and Nimaris pragmatically work together in the Andolan; several Nimari activists are highly regarded by the adivasis. Many Patidars can speak Bhilali because of their experience in dealing with adivasi labourers. Despite taboos about untouchability and food, adivasis and Patidars lived together amicably during the month-long Sangharsh Yatra. The only scornful note that I heard was struck by an adivasi who attended the planning sessions before the Yatra. He came back and announced that he was bored to death; the Patidars spent an hour wrangling over whether morning tea would be made during the Yatra! His comment was: 'We adivasis can go a lifetime without tea.' Many disparaging jokes were made about how the adivasis

---

[13] In part, this is due to the mediating presence of activists in Maharashtra and Alirajpur who are trusted by the hill adivasis for their demonstrated partisanship, but who also work harmoniously with activists from Nimar. Overarching unity is also created by Medha Patkar.

could walk forever, but the soft Patidars would not be able to survive the march to the dam site.

Despite co-operation between the Patidars and the adivasis in the Andolan, conflicts of class interests cannot be denied. Plains landowners employ daily wage labourers, mostly adivasis, who are paid twelve to fifteen rupees every day, even though the legal minimum wage for agricultural labour is Rs 25. Landowners aver that, if they paid minimum wages, their operations would become unviable. A couple of years ago, a conscientious Sub-Divisional Magistrate was appointed to the area, who vigorously started enforcing minimum wage regulations by prosecuting violators. His efforts were anathema to the landowners. Patidar politicking, using many of the organizational skills learnt through the Andolan, was instrumental in getting the official transferred. Social justice was temporarily sidelined.

Also sidelined in Nimar is the other issue raised by the Andolan: ecological sustainability. While the dam has rightly come to symbolize unsustainable and inegalitarian development, agriculture in Nimar is not based upon sustainable practices either. As described before, cultivation in Nimar is built around the use of machinery, synthetic fertilizers and pesticides, energy-intensive irrigation — the precise model of agriculture that is decried by environmentalists as the epitome of unsustainability. Thus Nimar is not an entirely successful embodiment of the Andolan's more general critique of development. As discussed in earlier chapters, the present circumstances of the hill adivasis are not sustainable either. However, their overall marginalization makes them highly credible bearers of the cause of social justice. But while the hill adivasis are absolutely deprived, a case cannot be made about the Patidars being downtrodden.

The Andolan has dealt with the anomaly of Nimar by using a two-fold strategy. First, it has showcased the hill adivasis — truly the worst-hit by the Project, and downplayed the presence of the Patidars. Thus, even though two-thirds of the population displaced by the dam lives in the plains, the plight of the hill adivasis, which personifies the theoretical arguments of the Andolan, is placed in the limelight. Second, the Andolan has focused on displacement and the injustice of it, setting aside questions about class and caste conflict in Nimar. Among the many layers of oppression, the Andolan concentrates on addressing the overarching cause of displacement, the injustice done to Nimaris by citydwellers and wealthy farmers in Gujarat, and ignores the conflict between classes *within* Nimar. Such strategic priority is, to a large extent, dictated by the exigencies of the battle. The urgency of opposing the dam requires that mobilization take place around the unifying issue of displacement. Other concerns are subordinated to this overriding one.

The Critique of Development and Creation of Alternatives

Potentially the most valuable task of the Narmada Bachao Andolan has been the bringing together of mass-based organizations from all over India. By forming a federation, these organizations have formally recognized their common agenda, and the importance of co-ordinated action. The Andolan is at the forefront of this coalition because of its initiative in making explicit the connections between particular grassroots movements and a broader analysis of development. This analysis consists of a critique of mainstream politics and includes a programme of action aimed at reform. This section will describe the Andolan's political critique and its attempts to create an alternative structure of action through a *jan andolan* (people's movement). It must be noted that this critique of development has been formulated by the activists in the movement and by supporters outside the valley; it is not the creation of people in the valley, both adivasis and non-adivasis, who understand the issue of displacement in a much more particularistic way.

The struggle in the Narmada valley has been linked to a general analysis of the class character of capitalist development, and of the appropriation of natural resources by a state which serves the interests of national elites and foreign capital (NBA 1991: 1–2). Large development projects are seen as symbols and synecdoches for the more gradual processes of commodification, resource-intensive industrialization and urbanization. These processes are believed to be informed by a *paradigm* of development which glorifies them as 'progress' and 'modernity'. While raising questions about social justice and ecological sustainability, the Andolan challenges this ideology of development which inspires projects such as Sardar Sarovar. Such a theoretical articulation of the agenda for the movement is especially favoured by the urban intelligentsia which supports the Andolan.

The political critique of the Andolan begins by deconstructing the agenda of *national* development used to legitimize the Project. According to S.C. Varma, former chairman of the Narmada Valley Development Agency:

No trauma could be more painful for a family than to get uprooted from a place where it has lived for generations . . . Yet the uprooting has to be done. Because the land occupied by the family is required for a development project which holds promise of progress and prosperity for the country and people in general. The family getting displaced thus makes a sacrifice . . . so that others may live in happiness and be economically better off (quoted in Alvares and Billorey 1987: 64).

The arrogance of assuming that people in the valley will willingly sacrifice themselves for the cause of 'the nation' is sometimes replaced by the argument that development is universally beneficial, that displacement is a change that people will welcome

in the name of progress, development and modernization . . . Why should any one oppose when tribal culture changes? A culture based on lower level of technology and quality of life is bound to give way to a culture with superior technology and higher quality of life. This is what we call development. What has happened to us is bound to happen to them because we are both parts of the same society (Joshi 1991: 68).

Here development is presented as an evolutionary process bringing about social homogenization that is universally desired and attainable.[14]

This is a classic statement of the way in which, what are in fact, particular interests are formulated and presented as general interests. To paraphrase James Scott, if the ideology [of development] is to become an effective instrument of consent, 'it must claim that the system of privilege, status and property it defends operates in the interest not only of elites but also of subordinate groups whose compliance and support is being elicited' (Scott 1985: 337). Such justification of the Project as serving a higher, generalized 'national' interest has been challenged by the Andolan, which argues that most of the benefits of the SSP accrue to dominant

[14] Ironically, Vidyut Joshi, the social scientist who has made this statement, is associated with a Gandhian institution in Ahmedabad — a sad commentary on the vast distance between the realities of tribal life and the ideas of even the relatively enlightened sections of the educated elite.

classes while its ecological and social costs disproportionately affect the poor. By pushing ahead with the Project, the Indian state has demonstrated that it promotes the interests of elites, represented as universal interests.

In the name of development, the Indian state safeguards the interests of elites by the repression of legitimate protest. The basic right to information about life-threatening issues is curtailed in the case of controversial projects such as the SSP. When popular consent is not forthcoming, the state tries to suppress even peaceful dissent through the use of terror. The Andolan has shown that the state, and the present political process of which it is part, is fundamentally undemocratic and violates the right of the people in the valley to participate in making decisions which critically impinge upon their lives. The violent response of the government to the Andolan has clearly revealed the true character of the state as elitist and authoritarian. The call of the Andolan for *Hamaare gaon mein hamaara raj* (Our rule in our villages) repudiates the legitimacy of the state and asserts in its stead the alternative of village self-government.

The slogan of 'Our rule in our villages' calls for non-cooperation with the state, the Gandhian method of passive resistance against exploitative authority. Jan andolan (people's movement), or decentralized and non-violent collective action, is posited as a political alternative to the dominant political system.[15] This is consonant with the Andolan's use of Gandhian tactics such as satyagraha, a strategy which seems to be deliberately chosen by the activists to ward off police action that could completely crush the movement. Of course, the use of Gandhian ideology is also motivated by an abiding faith in the moral force of non-violent action even though, as Medha Patkar once tiredly remarked, 'the days of moral pressure are gone'. The insistence of the activists on the use of non-violent tactics, which is in keeping with the wishes of the majority in the valley, has had to override some people's impatient urge to have a direct confrontation with the state.

Despite the disgruntlement of the minority, Gandhian tactics of satyagraha have prevailed. This is evident in the large, peaceful

[15] During the 1989 Harsud rally, too, the Andolan emphasized its distance from mainstream politics by refusing to allow Maneka Gandhi, the opposition politician who is active on environmental issues, to address the people.

demonstrations of the Andolan, often coinciding with the birth and death anniversaries of Mahatma Gandhi, and the moral pressure invoked through fasting, most notably during the Sangharsh Yatra and the June 1993 Bombay dharna. Civil disobedience has taken the form of non-cooperation with Project officials. However, while village self-government is theoretically consistent as a form of decentralized collective action that tries to create a political alternative to mainstream politics, it has met with an ambivalent response in the valley, especially in Nimar. People who support the Andolan are not always willing to boycott the government. Many of them are taken aback by the enormity of what is contemplated; for instance, if they stopped paying utility bills to the state, their water and electricity would be cut off and they would find it very hard to manage. In the event, the programme of 'Our rule in our villages' ended up being observed only nominally in Nimar; people symbolically boycotted census operations and state elections in a few villages. The limited success of this attempt to pressure the state while creating an alternative political process showed that the moment for village autarky has not yet come.

There is continuous debate within the Andolan about the merits of satyagraha versus more militant struggle, and about satyagraha versus mainstream electoral politics. The urgency of acting against the dam has meant that the Andolan has found it hard to distance itself from the structures of electoral politics. Lobbying, endorsing candidates from major political parties who declare their sympathy for the Andolan, and soliciting their intervention — all these are tactics that the Andolan has had to employ while trying to retain its character as a movement based on mass action. On the one hand, the Andolan has seized the moral high ground through its conduct of collective resistance outside the structures of electoral politics. On the other hand, it has had to set aside its ideological position and act pragmatically. Thus, the Gandhian slogan of village self-government is deployed pragmatically as one among several strategies to affect the state. The Andolan tries to exploit all available political spaces by simultaneously battling on many fronts — lobbying and pressuring state and central governments directly and through the intercession of sympathetic party politicians. Such pragmatism has not gone uncriticized; when activists from the Andolan went to Washington in October 1989 to present their case before the US Congressional

Sub-committee on Natural Resources, Agriculture Research and Environment in order that the Committee influence the World Bank to withdraw funding the Project, there was dissension within the Andolan about the ethics of asking a foreign government to intervene on an issue internal to India. However, since the United States is the World Bank's largest donor, the majority opinion within the Andolan felt that it was an entirely appropriate tactic to influence the Bank.

The political audience towards which the Andolan directs its collective action is composed not only of the Indian state and funding agencies, but also the elite 'reference publics' of the state — influential 'public opinion' in the cities (Lipsky 1968). Because pressure on the state is applied indirectly as well as directly, the Andolan has to orchestrate events of public protest towards three sets of people: the state, its 'reference publics' and the people. When people in the valley come together in a demonstration, the release of collective effervescence by the mass of singing, chanting bodies strengthens their resolve to fight. At the same time, the event is aimed, via the press, at crystallizing the views of an anonymous public — the literate middle class and above (the Andolan is not mentioned on the government-controlled radio or television). The need to gain the support of this fickle class compels the Andolan towards attention-grabbing, colourful events — the pledges against the magnificent backdrop of the river, for instance. When the building of an 'alternative political culture' requires years of patient work, the dam requires an instant Andolan.

Another problematic issue has been the structure of decision-making within the Andolan, a structure that is at odds with its claims to being a jan andolan. Like the Sangath in Alirajpur, the Andolan is also run by a group of outside activists who provide resources that local people may not have, and whose presence alters their perception of the risks of collective struggle. In addition, activists in Nimar have been instrumental in ensuring that the Andolan transcends the divisions of traditional party politics. The larger scale of operations of the Andolan means that activists have more tasks to shoulder — from speaking to Diet members in Japan to holding village meetings in the valley. Even though the Andolan's local activists in Nimar are literate, experienced in state level politics, and much more equipped to deal with political action against the state than their counterparts in the hills, they still seem to prefer a division of labour which leaves organization in the hands of the outside activists. Their attitude towards the outside activists seems to be, 'You run the Andolan; we will contribute our mite'.

Thus, the Andolan's struggle against the displacement of people in the valley by the dam, which is a completely just and worthwhile struggle in itself, fits somewhat awkwardly into its general critique of development. The mobilization in the Narmada valley has primarily been among people who will lose their land; while the case of the hill adivasis is perfectly apposite in illustrating the fundamental iniquity of the dam, the representation of the Patidars as socially deprived is much harder to make. At the same time, while the Andolan correctly criticizes the dam as environmentally disastrous, it begs the question of the ecological sustainability of current natural resource use practices in the valley, in the hills and especially in the plains of Nimar. The Andolan has chosen to accord priority to the cause of resisting the dam, arguing that it is the most drastic and immediately life-threatening manifestation of development. The urgency of acting against the Project has necessitated that the Andolan compromise to some extent with its goal of charting a course as a movement creating an alternative political culture based on Gandhian principles. What is evident at the grassroots is not a theoretically consistent, ideologically coherent movement, but a creative use of all available opportunities under highly adverse conditions. Such a response to the Project is forged by the activists who work with two disparate constituencies, uniting them to their mutual advantage. This gives rise to contradictions as well as fruitful conjunctions, both of which aspects are acknowledged by the activists.

The struggle in the valley, despite its rich complexity, has been understood and appropriated in quite another way by the urban-based intelligentsia who are concerned with representing the Andolan as fitting into a theoretical critique of the paradigm of development. This has led to their ironing out the awkward parts of the movement — the presence of the Patidars, for instance, or the absence of ecological sustainability — in order to demonstrate that the movement constitutes a theoretically satisfying challenge to the developmental state, even though the reality in the valley is more ambiguous. However, as mentioned earlier, the activists of the Andolan accede to this representation, collaborating in the showcasing of the hill adivasis and the simultaneous downplaying of the Patidars. In keeping with their strategy of using all available political spaces, activists feel that there is no conflict if the Andolan gains the support of a constituency which is primarily interested in using the movement to fulfil its own theoretical agenda.

However, in assimilating the resistance in the valley into its theoretical framework, the intelligentsia obliterates the mediating role played by the activists who facilitate the translation of particular interests into the general. The formulation of a critique of development is not a concern of the people in the valley; they are fighting as they have always fought — against outside oppression. If development is the legitimating ideology of capitalist accumulation — its representation — then the struggles of adivasis are made into counter-representations of resistance to development — a paradigm that most of them do not know or care about. Such appropriation, however politically effective in securing the support of the intelligentsia, results in the reification of the grassroots, a reification that in many ways is theoretically and practically problematic. The problems of persisting with a theoretical framework which simplifies politics and political agency into 'development' and 'resistance' are discussed in the final chapter. I shall conclude by attempting to present a more nuanced version of adivasi politics, one which restores to theoretical representation the contradictions inherent in the political process, and shall indicate how these contradictions may be resolved in practice.

10

Conclusions

. . . the Indian environmental debate is an argument in the cities about what is happening in the countryside (Guha 1992: 57).

The previous chapters have described the lives of Bhilala adivasis in the village of Anjanvara, their relationship with nature, and their resistance against the state, as expressed through everyday actions and beliefs. In this, the concluding chapter, I shall discuss the extent to which the struggles of adivasis fit into the general model of development and its alternative, as proposed by intellectuals theorizing on the subject. I shall begin with a recapitulation of the argument in the preceding chapters, then examine adivasi ideology and practice in the light of claims made about them in the literature on environmental movements. This will be followed by an analysis of adivasi consciousness and its representation by those who speak on their behalf, and the problems and possibilities engendered in the process of coming together. Finally, I shall discuss the expectations with which I had started research, and how they were challenged during fieldwork, leading me towards a different, more nuanced, understanding of adivasi politics and our role within it.

Summary of the Discussion

In the global march of development, the wealth of the earth is being appropriated by elites, impoverishing nature as well as the vast human masses who depend on natural resources for sustenance. The challenge to development has come in the form of political movements of people who are ecologically, economically and culturally marginalized. It is believed that the culture of 'indigenous' communities epitomizes a critique of the values underlying development, and that the beliefs and practices of adivasis constitute an alternative vision of an ecologically sustainable, socially just world — a vision that has inspired their struggle.

Over the centuries, adivasis have constantly fought an unequal battle against outside oppressors — the state and the market. Although power changed hands over time, being wrested from the Marathas by the Mughals, from the Mughals by the British, and from the British by the Indian nationalists, the adivasis experienced only a steady erosion of their material base and their cultural autonomy. National independence and the new project of development did not significantly alter this process; the universalizing claims of development's benefits were meaningless to people who did not identify with the concept of 'India' or 'Our Independent State'. Their lives continued to be ruled by bazaarias — bureaucrats and traders — as their resources were alienated by the state; they were compelled to enter into market relations and live under the hegemony of Hindu caste ideology. Resistance to this domination took the form of frequent uprisings, when adivasis would swoop down from the hills to which they were driven, and attack the villages of the plains. These adivasis were bandits whose raids were a constant trial to colonial powers engaged in maintaining 'law and order', but their ability to challenge the overall authority of colonial power was limited by their circumstances. Their tendency to importune their rulers for just settlements and their willingness to present injustice as *grievances* before the state, indicated that, however grudging it might have been, the adivasis accepted colonial domination.

The condition of domination has shaped the present-day lives of adivasis in the hills. They have carved out an identity and an existence which distinguishes them from their counterparts in the plains, even as they have felt the tug of the Hindu mainstream. Their isolation in the forested hills, relatively distant from centres of power, has enabled them to maintain a distinct language, religion and material culture which sets them apart from Hindus, but they *have* been influenced by Hindu values of caste hierarchy. Adivasi identity is expressed in the unity of the village community, defined by the clan, which stretches back in time to include previous generations, and which inhabits a particular physical space. The village is remarkably egalitarian in terms of land owner-ship, and forest resources are held in common. Households pool together their strength for labour-intensive tasks; and there is no differentiation between landowners and labourers. The com-munity is ruled by the ideology of reciprocity. However, the norm of generosity towards one's kin contains within it calculation and conflict, as shown in the arrangements of marriage. The clan comes together to negotiate the purchase of a bride, guided by the desire to drive a hard bargain and, at the same time, buttress its power and prestige. The rites of wedding symbolize the cor-porate unity of the clan as it acquires another female to carry forth its name.

The village community is defined in relation to the physical site upon which it subsists. Its economy is based on an agricultural cycle revolving around the use of the land, livestock, forest and river — a dependence that is acknowledged and secured through rituals of worship. Yet the apparent continuity of economic life is being undermined by ecological deterioration. The alienation of the forest by the state forced the confinement of adivasi cul-tivation to friable hill slopes, gradually eroding people's livelihood. At the same time, nevad — illegal adivasi fields in the forest — are the source of a steady stream of income for corrupt officials who extort bribes in order to look the other way. Today, the state justifies its continued ownership of the forest with the need for environmental conservation. However, its previous destruction of natural resources has so exhausted the environment that even the modest demands of impoverished people trying to create a living end up depleting natural resources beyond repair. The slow decline of the relatively self-sufficient hill economy has led people to be drawn unwillingly into the ever-widening circles of commodification of their produce as well as their labour. Constrained by their lack of control over their resource base, people lead contradictory lives, mining their future for the present.

At the level of their beliefs about nature, too, adivasi understanding is contradictory. On the one hand, their reverence for nature suffuses their everyday lives; they make strenuous efforts to secure the co-operation of nature through rites of propitiation. But respect for nature, whose uncertainties rule their fate, does not translate into a set of sustainable resource use practices. Beliefs about nature do not address ecological degradation — a concern that people living on the very edge of survival cannot entertain due to their lack of resources. While impoverishment and powerlessness compel people to carry on as usual, leaving little room for action that will deal with environmental deterioration, inaction is *also* in part a consequence of the shortcomings of traditional beliefs and practices in the present-day context. People think that the forest will always be there, and that it will regenerate itself. Or that the spirits that control nature can be appeased through sacrifice — a belief that acknowledges the power of nature over humans (a humility that is a welcome contrast from the hubris of industrialism), but which is not equivalent to an understanding of how the problems of degradation or disease should be managed. In the context of depleted natural resources, reverence is not enough.

The relationship between nature, adivasis and the developmental state is also expressed through more secular collective action. But besides political action against the state, people also engage in feuds, where the village community mobilizes to defend its honour against other villages. Feuds are something of an analytical anomaly for they are not about matters that fit into a theory of politics structured around development and 'resistance'. When adivasis fight each other, they are usually inspired by concerns such as the maintenance of caste taboos, the avenging of drunken insults and the management of recalcitrant women. The values guiding feuds —

patriarchal honour and caste purity — indicate that adivasi politics does not always embody the principles of progressive thought. These frequent conflicts show that the adivasi community is not an idyll of harmony and co-operation, but is lived as much through dissent and friction. For such untidiness, perhaps, this level of action tends to be ignored by scholars interested in assimilating adivasis into the politics of resistance. However, these concerns are held to be highly important by villagers, all of whom participate actively in their pursuit. Most significantly, women can use the dominant values of local politics to gain a degree of freedom denied to them otherwise; by eloping with lovers, they can defy the power of husbands and fathers over their lives.

The notion of 'community' is transformed by the work of the Sangath, seeking to bring together people as adivasis, united against the state, fighting to secure access to the land and the forest. The activists of the Sangath are important as mediators who provide resources that adivasis did not have before, and as agents engaged in the task of changing people's consciousness, creating the conditions for self-reliant resistance. The Sangath's experience of collective action has been mixed: although its members have succeeded in staving off the state's attempts to take over nevad fields, and have established a clout that discourages local officials from corruption, their agitational politics has prevented them from effectively taking over the institutions of local government. Their agenda has been limited both by the power of the state — repressive as well as bureaucratic, and by the dearth of resources that support them. However, through its links with similar organizations in the region, the Sangath has harnessed local struggles into a more general analysis of the conflict over natural resources and democratic rights, between adivasis and the state.

The argument incorporating local struggles into larger movements, both practically and theoretically, has been amplified further by the Andolan, which has focused even more explicitly on a critique of development, fighting against the injustice of the state's appropriation of natural resources from people in the Narmada valley. The Andolan has successfully united the disparate constituencies of Bhil and Bhilala hill adivasis and prosperous Patidar (upper caste) landowners in Nimar, transcending caste, class and party affiliations, and forging a common front against the Sardar Sarovar Project. The popular movement in the valley among people threatened with displacement, has attracted worldwide attention and support, leading to additional criticism of the Project on grounds other than displacement. For its critique of development and the 'national state', and for its attempt at creating an 'alternative political culture', the Andolan has been embraced by the intelligentsia.

However, the attempt to cast the movement in the valley as representing a critique of development disregards the class basis of the Andolan in Nimar, and the dominant presence of Patidars whose mode of production is the antithesis of sustainability and social justice. The people in the valley — hill adivasis as well as Patidars, continue fighting for their land, unconcerned by the niceties of theoretical frameworks and abstract political analyses. This has led to debates within the Andolan, between the movement in the valley and its urban-based supporters about ideological coherence and the long-term perspective of the Andolan.

The attempt of the Andolan to challenge liberal democracy through mass mobilization aimed at creating an 'alternative political culture' has also been somewhat difficult. The Andolan holds the state apparatus to be illegitimate and repressive; it tries to repudiate dominant political values through the moral pressure of passive resistance. But the urgency of stopping the dam necessitates that *all* kinds of strategies be adopted, including endorsing candidates from mainstream political parties whose ideology may contradict the espoused goals of the movement. While the Andolan asserts the establishment of an alternative state structure — village self rule — based on participation and decentralized power, the need to achieve rapid results has compelled the activists to temporarily set aside these stances for more pragmatic action, which is more in keeping with the realities of dependence on the state in the villages of Nimar. In order to fight against the fast-rising dam, the Andolan has brought together many disparate groups and ideologies, knitting them together in a political coalition that has been remarkably effective.

The three levels of politics — local, Sangath and Andolan — together show the possibilities as well as limitations in the ways in which the collective action of hill adivasis fits into a structure of development and 'resistance'. While people's experience of development has certainly placed them in opposition to the developmental state, the movement from objective conditions for opposition, to an ability to articulate an alternative vision of ecologically sustainable life, achieved through 'alternative politics', has been more difficult. How closely do adivasis resemble their portrayal as sustainable resource managers? To what extent can their politics be projected as an environmental movement? To answer these questions, we must begin by examining the different interpretations of the notion of 'environment', the meanings that it holds for different people.

Interpretations of 'Environment': Conflicting Perspectives

It has been argued that the appropriation of nature and labour was intrinsic to industrial development — an inherently resource-intensive and socially inegalitarian process. Therefore the struggle against development involves people marginalized by its processes. This is quite different from the way in which the relationship between development and environment is understood by national political elites. In India, environmentalists always tend to be depicted as 'boys and girls' — irresponsible and immature. Vidyut Joshi repeats this tactic by calling Sangath activists 'youngsters'; never mind that the average age of the 'youngsters' is more than thirty! (Joshi 1991: 70). In a conference on the Narmada conflict held in New York in March 1992, pro-Narmada bureaucrats persistently patronized Andolan activists by referring to them as 'our young friends'. In the words of Amarsinh Chaudhary, former Chief Minister of Gujarat, environmentalists are 'boys and girls' interested in saving 'tigers and trees'. The Gujarat government has consistently refused to even acknowledge the existence of the Andolan and the widespread movement against displacement in the valley, merely reiterating that criticism of the Project was limited to 'a few environmentalists', and could therefore be dismissed. The environmentalists have been further marginalized and referred to as 'ecofundamentalists' (Sheth 1991: 73). Such an understanding of what is 'environmental' echoes the thoughts of those who see development as an essentially benign process, marred only by a few regrettable externalities such as environmental pollution. 'Tigers and trees' are perceived as trivial concerns, luxuries that elites can afford to indulge in, since they have already gained the benefits of development. This interpretation of the conflict as 'environment versus development' has tended to prevail in government discourse.

The Sangath and the Andolan have, on the other hand, stressed the inseparability of ecological sustainability and social justice. Development is destructive because it worsens social distribution by reallocating resources from the poor to the rich. The Andolan's critique of the dam highlights its environmental dimensions: its impact on upstream and downstream ecosystems, and on soils in the command area, yet these effects are understood in terms of their human consequences. The Sangath's politics, which deals directly with the issue of adivasi control over the land and forest, is also based on this analysis of development. The Sangath and the Andolan's understanding of environmental conflict can be summed up thus:

Social movements of poor people are very often struggles for livelihood and they are ecological (whatever the idiom in which they express themselves) insofar as they express objectives in terms of ecological requirements for life . . . They are also ecological in that . . . they attempt to take environmental resources out of the economic sphere, out of the generalized market system . . . (J. Martinez-Alier, quoted in Watts n.d.: 23).

Omvedt also agrees that 'the new green movements in India have been survival movements of the rural poor' (Omvedt 1987: 36). Often these movements are uncomfortable with scholarly attempts to confine them to theoretical pigeonholes. Omvedt mentions that rural movements in India are wary of being labelled 'feminist' or 'environmentalist'. We note that there is a difference between people's perceptions of what they are fighting for — basic subsistence denied by the state, and the claims made by intellectuals who postulate that 'indigenous' resistance is a comprehensive critique of development based on the 'traditional' adivasi way of life, distinguished by its reverence for nature and simplicity — values that challenge the dominant worldview's desire for mastery over nature and material wealth. Although the ideology that perceives environmental conflict in terms of sustainability is external to adivasi consciousness, it is employed strategically by the movement in the valley to gain the sympathy of urban supporters.[1]

Being and Ideology

The complexity of ideology — the mingling of the public and the private, claims about sustainability and claims about subsistence — can be best unravelled by *locating* ideology, distinguishing between the people in the Narmada valley and their intellectual supporters. The ideology of development and resistance cannot 'sail through history innocent of any references to real individuals and the lives they lead . . . ' (Sayer 1987: 95). It is adivasis — not agents in the abstract, but definite socially and historically located individuals — who have ideas, whose subjective understanding guides their collective action and gives it meaning, and shapes their world. We cannot simultaneously hail adivasis as agents of history while dismissing their consciousness and their everyday lives.

In the final analysis, the relationship between social being and social consciousness can only be elucidated historically, over time, through empirical investigation of — in Marx's words — exactly

[1] Guha notes a similar tendency in the case of the Chipko movement which has two different faces — one 'public' and the other 'private'. According to Guha, although Chipko is not an environmental movement and is best described as peasant resistance, it has gained popular legitimacy by its use of the 'formal ideology' of environmentalism and Gandhian satyagraha (Guha 1989a: 173–7).

how 'people make their own history, but not of their own free will; not under circumstances they themselves have chosen but under the given and inherited circumstances with which they are confronted' (Marx 1852: 103). Ignoring how the historical circumstance of domination constrains consciousness leads to the misrepresentation of adivasis by intellectuals. In representing adivasis as living sustainably, and acting politically, inspired by a critique of development based on their 'traditional' values, intellectuals tend to romanticize subaltern forms of experience and culture, 'granting them a heroism that makes it difficult to understand "unheroic" decades' (Roseberry 1989: 46). Roseberry argues that if we make much too direct a connection between people's experience and the meanings that *we* feel they must attribute to it, we ignore the 'political implications of cultural inscription, the separation of meaning and experience in the context of domination'. Also, we ignore 'the ambiguity and contradictory nature of experience itself, an ambiguity that can only produce a contradictory consciousness' (Roseberry 1989: 46). Thus, we cannot automatically 'read off', or read into, the everyday experiences of adivasi life an ideology that is derived from an external critique of development.

At issue here is the difference between the beliefs and practices of adivasis and of those who claim to speak on their behalf. Instead of assuming a congruence between these two sets of ideologies, we have to explore how differences may be united in a synthesis which gains from the normative vision of the intellectuals and, at the same time, incorporates a more realistic view of adivasi life. Intellectuals are, after all, also agents of change, and their words carry more weight in certain settings. By inhabiting a particular social moment which is shaped by traditions of critical discourse, intellectuals bring to present struggles a historical richness — the mental production of previous generations of resisters and intellectuals (Gouldner 1979). They have provided the theoretical framework to link the social movement in the Narmada valley with those elsewhere in the world. Yet their eloquent championing of adivasi resistance has tended to obscure some of the difficulties that adivasis encounter in their attempt at resistance.

As earlier chapters described, adivasis in the Narmada valley inhabit a forest that has been considerably depleted by state-sponsored deforestation. The local pressure of using a reduced and degraded forest has led to further deterioration. State alienation caused the collapse of local institutions such as those dealing with common property that may have regulated the use of scarce natural resources. At the same time, the lack of security of tenure on nevad lands discourages the use of conservation measures. Soil erosion and deforestation by adivasis then takes place because they 'rationally' exploit their physical environment in an all-out effort to survive, an effort made all the more desperate because of a reproduction squeeze as terms of trade move against peasant producers, compelling them to exploit themselves and their environment even further. Thus, environmental degradation is embedded in overarching political and economic structures that are hostile to sustainable resource use by adivasis.

Under these circumstances, there is a contradiction between adivasis' reverence for nature and their unsustainable resource use. The only limited claim that we *can* make on their behalf is to assert that adivasis are 'environmentalists by default'. That is, their present resource use can be called sustainable only if we compare it with the vastly more destructive practices of the state and the market. In relation to the ecological devastation wreaked by the state in the Narmada valley: first deforestation, and now the ultimate solution of drowning the land — adivasis' use of resources appears miniscule in its impact. Thus adivasis are not sustainable resource managers, but in this grossly imperfect world, they come closest to that ideal.

But by ironing out the imperfections, not only in resource use practices, but also in the easy invocation of 'community' joined in an 'alternative political culture' in opposition to development, intellectuals allow themselves to be 'blinded by the glare of a perfect and immaculate consciousness' (R. Guha 1988: 84). Committed inflexibly to the notion of resistance as a generalized movement against development, they underestimate the power of the brakes put on it by the circumstance of domination.

To paraphrase Ranajit Guha, the objective of intellectuals and scholars of social movements is to take the history of insurgency from the continuum which is the progressive march of development and rearrange it along an alternative axis of a campaign for 'sustainability and social justice'. However, this too amounts to an act of appropriation which excludes the adivasis as the conscious subjects of their own history and incorporates them as only an element in another history with another subject. Just as it is not adivasis but development which is the real subject of these theories, so is an abstraction called 'indigenous peasant-tribal' — an ideal — made to replace the real historical personality (R. Guha 1988: 77).

For once a peasant rebellion has been assimilated to the career of . . . the Nation or the People [or against Development], it becomes easy for the [scholar] to abdicate the responsibility [s]he has of exploring and describing the consciousness specific to that rebellion and be content to ascribe to it a transcendental consciousness. In operative terms, this means denying a will to the mass of the rebels themselves and representing them merely as instruments of some other will (R. Guha 1988: 83).

When scholars such as Shiva and Bandyopadhyay claim that the practices of adivasis exemplify 'the life-enhancing paradigm' that has survived over centuries because adivasis have learnt to 'be like the forest, sustaining both the forest and the culture through time', they attribute to adivasis an environmental consciousness which is said to be ingrained in their 'traditional wisdom'. This claim derives from the scholarly agenda of setting up an ideological counterpoint to development. However, in trying to demonstrate that the critique of development actually exists in the lives of adivasis, intellectuals end up creating caricatures.[2]

The Politics of Representation

The privileging of elite consciousness which divides the world into

[2] The assimilation of adivasis into different ideological projects parallels the way in which the East came to be defined in different Orientalist constructions. A romantic and essentially positive view of the East is a mirror image of the scientific and essentially pejorative view normally upheld by western scholars of the Orient. In both cases, the East constitutes the Other, it is defined by a uniquely spiritual, nonrational 'essence', even though this essence is valorized quite differently by the two schools. Eastern people exhibit a spiritual dependence on nature — which, on the one hand, is symptomatic of their prescientific and backward self, and on the other, of their ecological consciousness and wisdom. Both views are monolithic, simplistic, and have the same effect — intended in one case, perhaps unintended in the other — of denying agency and reason to the East. 'The two apparently opposed perspectives have then a common underlying structure of discourse in which the East merely serves as a vehicle for Western projections' (Guha 1989b: 77). The position of Indian environmentalists such as Shiva and Bandyopadhyay repeats this structural dichotomy.

development and resistance does endow the struggles over land, forest and river with legitimacy in the eyes of environmentalists elsewhere. And, by linking local struggles to a global context, such appropriation *is* strategically important. Why does the image of adivasi resonate so powerfully in certain minds? The image has come to symbolize a normative vision of ecological wisdom — an inspiring quality for environmentalists in today's world, searching for cultures that embody a more respectful way of living with nature. But, however noble the cause, appropriation leads to the mediation of the adivasi consciousness by that of the scholar. The discourse of the general theory of development does not allow people to speak for themselves; it tends to be deaf to people's own understanding of their predicament. This is problematic because it slights the other, equally valid concerns of adivasis which answer to a different logic — of patriarchy, caste, honour. These areas of politics which are autonomous from development tend to be marginalized, even though, ironically, they come closest to constituting truly 'indigenous' 'alternative political culture'.

Intellectuals speak of the hegemony of development and how its myth of universal gain and progress, in which everyone believed for a time, has been shattered (Parajuli 1991). However, if development was the God that failed, it was never an adivasi God. People were never enchanted by the myth of development; how could they be when they only experienced its crushing exploitation? There could be no disillusionment when people had no illusions in the first place. The assimilation of adivasi struggles into an anti-development agenda neglects history — that people have always fought against outside oppression, on their own terms. Their history of resistance long precedes the advent of development.

The present theoretical treatment of adivasis reifies 'the grassroots', and is an idealization of people's actual life, a representation that is vulnerable to refutation. Their low-impact use of nature in earlier time was probably as much adventitious as it may have been deliberate; adivasis *were* limited by demography and technology from using resources destructively. Therefore it becomes hard to say whether their 'traditions' can be uncritically extolled as epitomizing sustainability, and what potential they hold as an ideal in the present, vastly changed, context. We cannot frame the adivasi past (or present) as a 'natural economy' — a starting point for a historical process that is a counterpoint to development, we have to come to terms with its disordered reality to create a more equal basis for co-operation.

Romanticizing adivasis reduces their problems and refuses to acknowledge that, at present, their ability to mount a critique has been vastly eroded by their subordination. While intellectuals as well as people in the valley stress that priority must be given to a need-based economy — a wholly sound basis for reorienting natural resource management, that, in itself, is not enough. The scale of the degradation of the land and forest requires a massive effort calling upon financial, technical and organizational resources — a magnitude that has been achieved so far only by the state. This leads us to seriously reconsider the strategic choices made by movements that want to stay clear of the sphere of party politics, moving perhaps towards a re-engagement in political struggles aimed at controlling state power, at least at the local level.

Idealization overstates the transformatory potential of adivasis acting in small, localized movements. It tends to downplay the power of dominant classes. It also underestimates the help and co-operation that is needed to challenge domination — which is a shame, since intellectuals, with all their resources, not the least of which is their commitment to the cause of fighting oppression, are so well situated to provide that help. Idealization ignores the role of the outside activists, whose presence empowers local peoples' struggles and transforms their consciousness. The activists mediate between adivasis and those who write about them; their perspective influences the intellectuals' view. This dialogue holds the potential for a fruitful coming together and crossing over.

Conclusions

The general argument about the theory of development and its environmental impact posits that it will be resisted in the form of environmental movements. This formulation, while perfectly valid in the abstract, becomes problematic when examined more closely in the context of a real set of conditions, those of the Narmada valley, and a real set of people, Bhilala adivasis. Glossing over the contradictions of people's lives is a tactic that prevents action towards their possible resolution. The treatment of the 'grassroots' as a 'pure' space for alternatives to development makes adivasis bear the burden of history. Our task cannot be simply that of rediscovery or excavation of a pristine 'indigenous' way of knowing the world, but must deal concretely with the problems in people's understanding and actions. If we do not respect people's understanding of what they are fighting for, we are being unfaithful to their history, their strengths and weaknesses, their truths. An appreciation of subjectivity is not simply an issue of representation, for representations are after all *regimes* that act upon the world. From an abstract, idealized expectation about adivasi politics, the environment and development, we come to a need to understand local struggles on their own terms, work to strengthen them and recognize their limitations.

In Chapter 2 I discussed the distinctive character of Indian environmentalism, where the Red project of changing the relations of production, and the Green project of using nature sustainably, are merged to create an ecological 'landscape of resistance'. As the case of Bhilala adivasis in the Narmada valley shows, the conflict over nature has several manifestations — from organizing to protect access to local forests to the world-renowned movement against Sardar Sarovar dam. The ability to mobilize across these different, yet connected, levels of action can be understood in terms of the relationship between local communities, activists and intellectuals — groups united in a common cause, yet embedded in different social contexts and moved by different ideologies.

In the case of the Narmada Bachao Andolan, fighting against the dam, and against destructive development in general, has necessitated the mobilization of a panoply of strategies which span the ideological spectrum of environmentalism. Critics of the dam employ arguments challenging the wisdom of large, capital-intensive projects, calling for the use of appropriate technology. The issue of the displacement of a vast population raises questions about the social distribution of costs and benefits, implicitly drawing upon an ecological Marxist understanding of the nature of development. At the same time, the attempt to engender an 'alternative political culture', opposed to the developmental state and mainstream politics, builds upon Gandhian traditions of decentralized and non-violent collective action. However, the urgency of fighting on all fronts has compelled the Andolan to reconcile constituencies and ideologies that are sometimes at odds, leading to an alliance that, despite its success, retains elements of unease.

The work of the Andolan complements that of the Sangath, which attempts to bring about a more deep-rooted transformation of consciousness. The Sangath's longer time horizon and more modest scale of operations allow it to engage more thoroughly in the task of organizing people against the state which has alienated their natural resources. While insurgent consciousness is inextricably tied to the issue of livelihood, and action against the state is motivated by the desire to safeguard the ecological basis of survival, popular mobilization requires a transmutation of identity. From a preoccupation with the politics of honour, people come together as adivasis, unified by their shared experience of exploitation. Thus, although the ideology of the Sangath is broadly ecological Marxist in its orientation, circumstances cause it to critically address the issue of identity based on cultural values other than class, thereby reshaping the ideology of ecological Marxism and moving beyond it.

These streams of ecological consciousness are joined together with yet another strand — the cultural traditions of adivasis, to form a powerful, visionary critique of development which, despite the contradictions embedded in its lived reality, promises to inspire environmental action in the future. The three aspects central to adivasi life — the gods that are nature, land and forest, community — hold the potential to form a challenge and an alternative to development when supported by the help of activists and intellectuals who listen and observe as much as they speak and write. It must be our task to transform our understanding into action and aid, and together forge a future that is more just — to people and to nature.

Epilogue

Involvement with a place and its people can endure long after the experience of fieldwork has been inscribed and interpreted in a scholarly article or book. After finishing my studies, I returned to the Narmada valley to work with the Sangath as an activist. I had initially decided that I would take the Sangath's education programme in hand: organize regular training sessions for the young men who taught in other villages, try to ensure that teachers taught regularly and got their wages on time from the village, print text books in Bhilali. However, my modest plans were scattered to the winds when a storm broke loose over Anjanvara, the village where I had lived. In this epilogue, I want to pay homage to the people of Anjanvara who courageously defied the Jhabua administration and stood by their resolve to not leave their land and gods. With all that they hold precious at stake, resistance, for the people of Anjanvara, could not be an impetuous act of brave folly, but consisted of carefully considered acts of courage.

September 1992: Bowing to persistent criticism and adverse publicity of its involvement in the Sardar Sarovar Project, the World Bank announced that it would reconsider its support for the SSP at the six-monthly review due in April 1993. A key factor in the Bank's decision was to be the performance of the Indian government's programme to rehabilitate villages in the first-phase submergence area. The record of the Madhya Pradesh administration came in for special criticism in the Bank's report, which pointed out that Resettlement and Rehabilitation (R & R) work in the state had been grossly unsatisfactory. 'While a good deal of data has been assembled, it is still inadequate for the purposes of R & R . . . Consultation with Project-Affected Persons (PAPs) has also not been adequate and planning . . . largely absent except for a continued assertion that PAPs can go to Gujarat . . . '[1]

January 1993:   In the first two weeks of the month, the Collector of Jhabua went on a *padyatra* to all the river-side villages in Alirajpur tehsil to 'inform people about rehabilitation'. This was a momentous event: never before had any District Collector ventured out to these remote villages. The Collector told the journalists who accompanied him that his mission was to bring the message of resettlement to the 'innocent and misguided tribals who were being incited by outsiders to oppose the dam'. Thus, a caravan of 150 people, a third of whom were policemen, marched along the river, presumably to fulfil the World Bank's demand for 'consultation with affected villagers'. In some villages, people stopped the Collector on the outskirts of the village and told him that they did not want to move. In other villages, people shouted anti-dam slogans and demanded that the Collector leave their land. In Kakarsila village, the Collector attempted to initiate a dialogue, saying that he had come not to talk about the dam, but to solve local health, education and drinking water problems. The village sarpanch gave a bitter laugh. 'Never in my lifetime nor in my father's lifetime have we ever seen a Collector', he said, 'and now you come to talk about schools and handpumps when you want to push us off our land!'

The collector's caravan left, but not before it had established a 'rehabilitation camp' in Sakarja village, where most of the families had agreed to move to Gujarat. From this camp, survey officials would, under heavy police escort, try to force their way into other villages along the river to carry out survey work and intimidate people into leaving. While surveys may seem like innocuous acts of data collection to outsiders, in the highly charged atmosphere of the submergence zone they took on an altogether different meaning. Since each earlier episode of surveying had been marked by brutal attacks on the villagers by the police, surveys had come to represent the invasion of the state into their lives (see Chapter 9). People living near the Sakarja camp complained that the government officials staying there had

---

[1] World Bank. 11 September 1992. *Review of Current Status and Next Steps Regarding SSP.*

defecated into the stream, fouling their only source of drinking water. The police at the camp would stop people travelling along the river and harass them for no reason. People had come to know, too, of a similar camp in Kakrana where villagers were terrorized by the police, who hoped to drive a wedge into their united opposition to the dam.

The physical presence of the Sakarja camp dominated the entire stretch of the Narmada. Seven gigantic tents of thick white canvas, with blue doors flapping in the breeze, were visible for miles on end. Sakarja is located on a bend, but set back from the bank so that none of its houses are visible from the river. The rehabilitation camp was set up on a high promontory overlooking the river and, as one walked upstream from Dubkheda to Anjanvara, the Sakarja encampment was an inescapably large presence on the landscape. At night, the roar of generators and the glare of powerful spotlights would cut through the darkness, violating the peace of the valley. The very sight of the tents came to symbolize the intrusive power of the state over people's lives: mocking people as they went about their business every day, daring them to defy the might of the government. After a meeting, the area's villagers decided that the people would ask the government to remove the camp which destroyed their peace. This was announced to the public and the press.

On 22 January 1993, about 150 people, mainly women and children, gathered in Sakarja and went in a procession to the campsite. After telling government officials that they would have to dismantle their tents and leave by sunset, people sat down on a peaceful dharna, singing songs and chanting slogans. The mood was festive, rather like a large picnic, and people sat there all through the day, taking turns to go and eat and drink. There were only a few bemused-looking government officials around. They were asked to remove their belongings from their tents, for if they did not dismantle the tents then the people would do so themselves. It was only just before sunset, however, that the by-now nervous officials went into the tents to move their most valued possessions: a television set and a VCR. Then, as the officials stood by and watched, people swarmed around the tents and untied the ropes that held the canvas to the poles. With great efficiency and swiftness, the poles were uprooted and the canvas laid on the ground, but care was taken to ensure that nothing, not even a tubelight, was broken. Some people gleefully debated whether they should round off their protest by urinating on the collapsed tents, but decided against it! Then they went down to the river, cheering all the way, leaving two policemen standing silhouetted forlorn against the setting sun, their tents gone. It was all over in ten exhilarating minutes.

29 January 1993: It was the day after Bahaduria's indal. Most of the guests had left, but a group of close relatives remained in Bahaduria's house, at the upstream end of Anjanvara, to eat the head of the goat that had been sacrificed the previous day. Early in the morning, at the downstream end of the village, as Dhedya and his son Vesta were releasing cattle and goats from their pens, they heard Dhedya's youngest son Vahria shout, 'The policemen are coming!' Leaving the lowing cattle with their ropes half undone, Dhedya rushed out and looked where Vahria was pointing. Coming up the path that came from the river was a long line of bazaarias, some in police uniforms and carrying rifles, some in civilian clothes, about sixty in all. Dhedya instantly knew that they must have something to do with the dam: that was the only thing the government ever came for.

Dhedya's house is the very first at the downstream end of Anjanvara, towards Sakarja. It was the first house that the surveyors had to measure. Dhedya told them that he did not want to move, and that he did not want his house and his land surveyed, but they paid no attention to him. An officer called out an order and two men unrolled the measuring tape, impatiently pushing aside Dhedya as he tried to stop them from reaching his house. A policeman with a lathi moved towards him menacingly and barked, '*Abe,* will you move or not? Or do I have to beat some sense into you?' Dhedya, about sixty five years old, was reduced to pleading ineffectually with the surveyors while they walked all over his property. Most of the men from his phalya were at Bahaduria's house, too far away to come to his aid. Then the police and the surveyors went on to the next house, that of Dhedya's son. Here, too, the people of the house were outnumbered by the police so that, against their wishes, their land and house were forcibly measured. The survey team traversed the village house to house, pushing away protesting villagers.

By the time the survey team reached the upstream end of the

village near Bahaduria's house, however, quite a crowd of villagers had collected. All of them were filled with the angry frustration of powerless spectators who had witnessed strangers trespassing on their land: strangers who threatened violence if they protested against this injustice. As the last few houses were being surveyed, these individual feelings of burning resentment joined together into a collective response to the threat. The surveyors were outside Bola's house when this anger was translated into action. Prompted by the desire to do *something, anything*, to stop the survey, Bola's daughter, fifteen year old Danki darted forward and grabbed the measuring tape out of the hands of a surveyor. 'We've said that we won't move', she said, 'so why do you come again and again to our village?' One of the policemen, the Station House Officer of Umrali town, thought that some of the people in the crowd had been at Sakarja on the day that the tents were dismantled. 'Why did you break our tents?' he demanded. 'You feel so bad because your tents were brought down', Danki replied indignantly, 'we are losing everything that we have and you don't care at all!'

The police officer swung his lathi and knocked Danki to the ground. As she lay there, he prodded her with a rifle butt. While the rest of the village watched from a few feet away, Danki's mother, Velbai, and her aunt, Singi, ran forward to save Danki. They were also beaten with rifle butts. Seeing their women being beaten was more than the villagers could bear. They picked up stones and started pelting the government team with them. The police precipitately retreated back to the river by the path along the patel's house, firing ten rounds in the air as they passed the *boor* tree by Khajan's house.

30 January 1993: The people of Anjanvara knew that the police would be back, this time thirsting for revenge, eager to teach a lesson to the adivasis who had dared to oppose them. Fearing the worst, but determined to avoid a violent confrontation, the villagers decided that they would all hide in the hills if the police returned. Only if the police captured a villager would they take action. Pairs of men were sent off to Attha and to Badwani to inform activists of the Sangath and the Andolan about the events of the previous day. Everyone kept a lookout along the river towards Sakarja. But the police came in two parties that afternoon, one along the river, and the other from the hills behind the village. More than a hundred men of the Special Armed Force entered Dhedya's phalya from both sides, trapping some women and children in between, who ran to the river in a panic. The police party which had come in through the hills entered houses in the village, which lay deserted since everyone was hiding in the hills. In the patel's phalya, they ransacked houses and vandalized whatever they could lay their hands on. Earthen water pots were smashed, metal utensils crushed, and poultry, groundnuts and money stolen in sixteen homes. An old muzzle-loading rifle belonging to Dhanya, the village patel, was confiscated as were bows and arrows from every home, presumably to 'prove' that these weapons had been used against them by the villagers.

The eight women from the downstream phalya, meanwhile, who had run to the river, were huddled behind the stark, hot stones on the river bank, trying to make themselves as small as possible. But the police spotted them and surrounded them. Dhedya's elderly wife Jasma, his sons' wives Rangi, Jemali, Lasmi and Visi, Bandi and Chhinki — his married daughters who had

come to visit, his neighbour Gulya's wife Hudki — with their
infants at their waist and the slightly older children clinging to
their sides, ran into the water, trying to swim away from the police.
They were dragged out and told that they would be taken to
Sakarja where the motla sahib (Collector) wanted to see them.
The protesting women and the children were prodded with lathis
and rifle butts to make them walk faster towards Sakarja. Children
who couldn't keep up with their mothers were struck from behind.
Rangi, who was five months pregnant, was repeatedly hit on her
belly and back with a rifle. When she cried out, 'Don't hit me! I
have a child in my belly', one of the policemen snorted, 'You're
fat because you eat too much. You deserve to be beaten.'

As the women and children were being led along the river
towards Sakarja, an anguished Dhedya and his sons Malsingh and
Vesta watched from the hillside. They ran along the hillside,
parallel to the departing police party, shouting at them to let the
women go. 'Your quarrel is not with these innocent women and
children', cried Dhedya, 'what harm have they ever done to you?
With folded hands, I beg you, let them go!' The police simply
prodded the women with their lathis and pushed the children
ahead to make them walk even faster. Seeing their women and
children in danger, at once desperately afraid for them and angered
at this challenge to their honour, the men on the hillside pelted
the police below with stones. The police responded by directing
seven rounds of .303 rifle fire at the villagers. Dhedya's son Vesta
was injured on his shoulder and hand. But the villagers kept
throwing stones, forcing the police to abandon the women and
beat a hasty retreat.

31 January 1993: On 30 January while Anjanvara was facing the
wrath of the administration, the full-time activists of the Sangath
were in the village of Vakner, concluding a group meeting. Among
other things, we discussed the plan of work for the following
months, including a set of training camps for village activists,
women and youth. As the meeting came to an end, we heard that
there had been police firing in Anjanvara during the survey.
Frantic with worry — had anyone been injured? perhaps someone
was dead? — we tried to figure out the best thing to do.

We had felt for some time that police pressure on the Alirajpur
part of the submergence zone had reached unprecedented heights.

Friends had told us, too, that the police planned to arrest the Sangath's activists in connection with the Sakarja tent incident. We learnt that the district administration had planted a fictitious version of the incident in the press, in which activists had burnt down tents, stolen valuable equipment and stoned the police. The administration had also fed stories to the local press alleging that the Andolan was losing its following in the Alirajpur villages because of the Collector's 'education' of the villagers, designed to 'guide' them along the 'right path'. The Andolan, this fiction alleged, was resorting to desperate and violent actions in order to cling to its support base. In order to counter this propaganda, we decided that we would go to Indore with villagers who would themselves describe to the press how the administration was trying to browbeat them into submission. Some activists were assigned the task of organizing this visit. We also knew that a padyatra of the Andolan would pass through Alirajpur in a few days; it would include Medha Patkar and prominent supporters of the Andolan from all over India who would hold a meeting in Anjanvara on 5 February. We sent word to them about the present crisis.

Two of us, however, left for Anjanvara to find out what had happened there. We could not go by the direct route since the Sakarja encampment straddled the path, so we took a roundabout route through the hills behind Sakarja, a tortuous climb on a narrow slippery path in the dark. After four hours of negotiating unfamiliar trails, we were well and truly lost. We slept in a deserted field, hungry and thirsty, frustrated at being so close to Anjanvara yet unable to reach it. The spotlights from the police camp at Sakarja were a bright glow in the sky. At night, we heard the rumble of jeeps in the distance: were more police reinforcements being brought in, or were the corpses of villagers being transported out, or was it merely the Collector visiting the scene of the incident? The next morning, we hazarded nightmarishly steep paths and, after three hours of following false trails and then retracing our steps, finally reached a house! A short rest and a meal and we were off again, this time with someone to show us the way, towards Anjanvara.

As we approached Dhedya's house, it became evident that something was horribly amiss. Dhedya and his sons were standing outside with their bows and arrows aimed at us which they only lowered when we shouted 'Zindabad!' from far away. Then we

were surrounded by all the people of Dhedya's house, all talking
at once, anger and relief spilling out in a confused tale. That was
when we learnt that the police had returned to Anjanvara the
previous day in even greater numbers and beaten up the women
and children of Dhedya's phalya. The women were still trauma-
tized by what they had gone through, and petrified that the police
would come back again and carry them off. Their children clung
to their sides, anxious to not let their mothers out of sight.
Rangibai, who was pregnant, kept crying, wondering about her
unborn child. Dhedya alternated between rage at what the police
had done and fear of retribution in the future. As we went
through the rest of the village, our outrage, too, grew. Every house
in the patel phalya had been systematically ransacked. Several
houses had nothing left in which to fetch or store water, no vessels
in which to cook. The houses of people who were known to be
active in the Sangath were singled out for special destruction. After
hearing what had happened, we decided that there was no time
to be lost; we would leave for Indore with a couple of people who
would describe what had happened to them.

Disaster consists of small, but crippling events. As we were
leaving the village, someone came running to say that Rangibai
was in great pain, perhaps she was miscarrying. One of us would
have to go to Badwani to arrange for a jeep to take her to the
doctor. Fortunately, two other Sangath activists reached the village
at just that moment, who could take care of organizing an improvised litter and four bearers from the village to carry Rangibai
to Kakrana, the nearest motorable point. After another long and
exhausting walk through the hills, we finally reached Kakrana late
at night. We then had to reach Dahi, twenty kilometres away,
where we could snatch three hours of sleep before the bus left.
We reached Indore the next afternoon and found other Sangath
members waiting for us. Among them was Khajan, who had left
Anjanvara a day early to inform the Badwani office of the Andolan
about the first police attack. When we told him that his house
had been ransacked the day that he had left for Badwani, Khajan
was quiet for a long time. Then he asked, 'Are my wife and
children all right?' On learning that his brother had escorted his
wife and their youngest child to her father's village Arda for safety,
Khajan looked a little reassured. We contacted our friends in the
press, told them about the events in Anjanvara and the collector's

campaign against us. We decided to hold a press conference on 2 February 1993.

2 February 1993: In a sensational swoop, the police arrested nine activists of the Khedut Mazdoor Chetna Sangath as they came out of the Indore Press Club at noon. The Jhabua Collector told the press that he was glad that 'long-wanted Naxalites had been captured at last'.

In Anjanvara, just as villagers were trying to pick up the pieces of their shattered lives, the police returned. All the villagers fled to the hills from where they watched the police and forest officials enter their homes. The police left after six hours. The villagers returned to find their houses utterly destroyed. In an act of calculated malice, the police had smashed twelve of the heavy grindstones that were essential for making rotis. Not only were the grindstones expensive, in a remote village like Anjanvara, they were almost irreplaceable. The police broke hearths, brought down fodder platforms and tore fishing nets. They smashed the few pots that had escaped their last visit, took away timber, and broke walls and tiles from the homes of village leaders. They ate, spoiled and stole crops of groundnut and chana. They found precious hoards of seed grain and threw them into the dust for hens to eat. Anjanvara is a poor village; the police inflicted damage worth over fifty thousand rupees.

The other nine people who had gone with me to Indore were arrested; I escaped because I left the Press Club a little before everyone else. I only learnt of the arrests in the evening when I started making inquiries after waiting in vain for several hours for the others to return. I also came to know that there were four non-bailable criminal cases against us; among other things, we were charged with section 307 Cr.P.C. — attempted murder. I caught the night bus to Badwani and woke up the Andolan activists with this news. The morning brought more bad news; two men came from Kakrana to announce that the police had arrested three men from their village. While the Andolan activists got in touch with their Baroda office, I caught the bus that would take me to a village close to Dahi, from where I would walk to Anjanvara. As I sat waiting for the bus to start, I noticed that the other passengers were giving me odd looks. The conductor, with whom I was familiar, came up to me. 'Madam, you people have kicked up quite a storm, I hear', he said. 'What have you heard?' I asked. 'Well, we have been told that your organization has been banned. We were warned that if we let you aboard our bus, the entire vehicle would be impounded', he said. 'It will be much better if you don't travel.'

So I asked to be let off the bus at Kadmal, close to Badwani. In Kadmal, various Andolan activists discussed this latest development. In fifteen minutes they had arranged for a motorcycle on which Jagdishbhai would take me towards Dahi. As we were setting off, someone pointed out that I would be very conspicuous on a motorcycle; what if I were recognized? I was sent off to Sitarambhai's house where his daughters enjoyed themselves hugely by dressing me up as a Nimari peasant woman. My knapsack was disguised with a fertilizer sack and we roared away, looking to the world like a Nimari couple on their way to the market.

After I was left in Dahi, I walked through Kakrana to Jhandana, telling villagers of the Indore arrests and of the forthcoming meeting in Anjanvara on 5 February. The next day, as I walked to Anjanvara, I met Sangath members from Bhitada and Sirkhadi going in the same direction, carrying flour for their relatives. Then I learnt that the police had returned to Anjanvara and that all the grindstones in the village were broken. I found the villagers subdued, surrounded by smashed possessions which they were still too stunned to clear up. My news that nine people, including Khajan, Vesta and Kaharia of Anjanvara, had been arrested in Indore didn't help matters any. But by evening, people from other villages had started arriving for the next day's meeting. They came in batches of ten or twenty, called out 'Zindabad!' and came to the fire to light their bidis for a smoke after their long journey. In the leaping firelight, we recognized faces and exclaimed our pleasure and asked after their families. From Toorkheda in Gujarat, from Sikka, Maal, Danel and Pola in Maharashtra, from Kakarsila, Jalsindhi and Dubkheda in Alirajpur, they kept coming all through the night. And the people of Anjanvara were roused into hectic activity, arranging food, firewood, water. After the loneliness and the worry of the last few days, there was tremendous comfort to be derived from this sharing and solidarity. This was the strength of the Sangath and the Andolan writ large.

5 February 1993: More than three hundred people attended the Anjanvara meeting. Besides villagers from the submergence zone, there was a human rights activists' team, journalists and film-makers, participants from the Narmada padyatra, and a group of activists from Nimar who had brought a jeep-load of grain and *dal* for Anjanvara. The police tried to dissuade people from reaching Anjanvara, but stopped short of arresting them. Medha Patkar was detained for several hours at Bakhatgarh but finally allowed to go on. We decided at the meeting that we would hold a rally in Alirajpur on 15 February to protest against police repression. The next ten days would be spent in mobilizing people for the rally. An Andolan activist from Delhi was sent to Alirajpur to meet our lawyer and to arrange for bail for those in jail.

6 February 1993: The Member of Parliament from Jhabua visited Anjanvara, talked to villagers and examined the devastation in their homes. Later he told the press that police excesses were aimed at crushing people's movements.

The Sakarja rehabilitation camp was dismantled by the ad-ministration.

My husband, the Andolan activist from Delhi, who went to Alirajpur to inquire about those in jail, was arrested within ten minutes of reaching the town. Like everyone else, he, too, was charged with attempted murder.

The next week, we were 'underground', according to the police. Meanwhile we held a press conference in Delhi, visited supporters in Indore and consulted our lawyer friends, sent off a First Infor-mation Report (FIR) about the Anjanvara incident, prepared banners and handbills for the 15 February rally, and got an-ticipatory bail. The Andolan machinery in Badwani worked over-time to arrange for bail guarantors, vehicles and people for the rally.

All sorts of news about events in Alirajpur filtered through to us. The Collector held a meeting in the town where he announced that the Sangath was a Naxalite organization that must be uprooted from the area. He urged the townspeople to socially boycott Sangath activists. Policemen told the landlord of the Sangath office that if he did not throw his tenants out, his entire house would be seized by the police. With the active support of the local administration, a rally against the Sangath was taken out on 9 February, demanding that activists be externed from the district. The police arrested more than twenty Sangath activists from all over Alirajpur, some from villages eighty kilometres away from Anjanvara. They were all charged with attempted murder.

The arrested activists were severely beaten in police custody. Khajan was tied to a pillar in the Alirajpur police station and kicked and beaten on the head till he fainted. Hengibai, the only woman arrested, was threatened with rape. All the men were handcuffed and paraded in the streets of Alirajpur while being poked and prodded with lathis. The administration had apparently decided to wage war against us 'Naxalites'.

Despite the administration's efforts, the 15 February rally in Alirajpur was a success. The police put up road blocks to prevent people from reaching, yet villagers walked more than seventy kilometres to come together and protest against the blatant violation of their right to organize and exercise their political will. The rousing slogans became louder still as we passed outside the jail where our comrades were held. Later, when they were out on bail, those arrested told us that the sounds of the rally were clearly heard inside the jail. Everyone came out into the courtyard and listened. The other prisoners were so impressed upon hearing that this was a Sangath rally that they offered scarce, prized bidis in admiration to Sangath members!

In March 1993, Khedut Mazdoor Chetna Sangath filed a writ petition in the Supreme Court of India about the Anjanvara and Kakrana incidents, and about the Jhabua administration's concerted attempts to crush an organization engaged in struggling for basic human rights. The Court was shown photographs of handcuffed Sangath activists being paraded in Alirajpur. Since the Supreme Court has issued clear directives which severely limit the circumstances under which undertrial prisoners can be handcuffed, the Court found prima facie evidence that the Jhabua administration stood in contempt of court. Criminal contempt proceedings were instituted against the Jhabua district Superintendent of Police and Alirajpur police officers. The Court has also ordered an inquiry by the Central Bureau of Investigation into the 'incidents of beating and torture in police custody'. The case is being heard before a bench headed by the Chief Justice of India. A couple of villagers came from Alirajpur to attend the hearings in Delhi. Although it was all in English, they still found the proceedings immensely satisfying. The men of Anjanvara now twirl their moustaches with challenging disdain when they come across a policeman in the haat; after all, didn't Khajan see senior police officers cringing in the motla court in Delhi? Sometimes small pleasures can be infinitely sweet.

Appendix 1
Trees in the Forest around Anjanvara and their Uses[1]

Aamba (Mangifera indica) — edible fruit, wood for making drums
Aambadu — edible leaves
Aamli (Tamarindus indica) — edible fruit and leaves, wood for home construction
Ali (Meyna laxiflora) — edible fruit
Anjan *(Hardwickia binata)* — fuel, fibre, wood for making cots and for home construction

Babulya (Acacia nilotica) — fodder
Baheda (Terminalia bellirica) — fruit used in Ayurvedic and Unani medicine
Bedu — fuel, seeds for oil, fodder for goats
Beela (Aegle marmelos) — leaves used in worship during kholo pooja
Beeya (Pterocarpus marsupium) — wood for home construction, fodder
Boor (Zizyphus jujuba) — edible fruit, fodder, wood for making ploughs and for home construction
Bufulu — edible fruit, colds are cured by waving leaves around the head, and then boiling them in water
Butko (Clerodendron roxburghii) [1908] — wood used for tying bullocks when threshing, for making drums

Charoli (Buchanania lanzan) — edible seed, fuelwood

[1] The names of the trees are in Bhilali, listed in alphabetical order. I have been able to give the botanical names of only a few of them. 1908 in parentheses after the botanical name indicates that the tree was identified thus in Luard 1908a.

Dhaavda (Anogeissus latifolia) — edible gum, wood for making fodder storage platforms

Ganyaru (Cochlospermum religiosum) — edible fruit
Girvala — fuelwood
Gurad (Milletia ovalifolia) — wood for home construction
Gurbuta — fuelwood, fodder

Haagda (Tectona grandis) — fuel, wood for drums, agricultural implements, kitchen utensils and home construction, leaves for packaging
Haalai (Boswellia serrata) — aromatic gum resin, fuelwood, fodder
Hagnia — fencing
Heglu (Moringa oliefera) — edible leaves, fodder
Hehtu — wood for home construction
Helti — edible leaves, fuelwood
Hemlu (Bombax ceiba) — edible flowers, wood floats on water
Hivni (Dalbergia sissoo) — wood for making agricultural implements and drums
Hiyali (Nyctanthes arbor-tristis) — wood used for making screens, fruit is fodder
Hoijadu — wood for home construction

Jaamun (Syzygium cumini) — edible fruit

Kadai (Sterculia urens) — edible gum
Kambu (Metrigyna parviflora) — wood burnt during the festival of Holi
Kanji (Pongamia pinnata) — twigs used for brushing teeth
Khairu (Acacia catechu) — fodder, wood for making pestles, heavy wood for construction
Koalu — wood for home construction
Kothar (Wrightia tinctoria) — fuelwood, fodder
Kudi (Holarrhena antidysenterica) — wood used for making ladles
Kuham (Schleichera trijuga) — branch worshipped during Indal, edible fruit and leaves, wood used for home construction
Kursilya — bark used as an ointment on cattle's wounds and sores
Kutarandu — wood used for home construction

Maandal — wood for agricultural implements
Manhing (Dolichandrona falcata)[1908] — fodder

Moyni (Odina wodier) [1908] — paste of fruit used to stupefy fish, wood for home construction, fuel

Muhda (Madhuca indica) — edible flowers, also fermented into liquor, fruit for oil, wood for home construction, fuel

Mukhu (Schreibera swietenioides) [1908] — fuelwood, edible gum and leaves

Neemda (Azadirachta indica) — fuel, wood for drums, twigs for brushing teeth, leaves are medicinal, edible fruit

Ningodi (Vitex negundo) — wood for making screens

Okan — fuelwood, fodder, edible fruit

Onaje (Annona squamosa) — edible fruit

Oyania — medicine for cuts, also if cattle get inflamed infections

Pahal (Butea monosperma) — leaves used for worship, also for cooking pannia, fuelwood

Peepur (Ficus religiosa) — fuel, edible fruit

Ragatrenu — potion made from the bark drunk for diarrhoea

Rekhul — thorny fencing, fodder

Rinjanu — leaves for fodder (fruit poisonous to goats)

Taad (Borassus flabellifer) — sap fermented into liquor

Tamblya — sore throats are cured by brushing teeth with twigs

Temru (Diospyros melanoxylon) — leaves used for making bidis, edible fruit, wood for making carts

Tidkya — backaches are said to be cured by fixing iron into the trunk of the tree

Tini (Ougeinia dalbergioides) — wood for making agricultural implements

Ulu — edible seeds, wood used for making drums, fuel

Umbar (Ficus glomerata) — edible fruit

Vaahan (Dendrocalamus strictus) — bamboo culms for weaving baskets and screens, making flutes, bows and arrows, for home construction

Vad (Ficus benghalensis) — wood for making utensils for a wedding

Appendix 2
Gayana — The Bhilala Song of Creation

As discussed in Chapter 7, the gayana is central to Bhilala cosmology. By singing the gayana during Indal pooja, people re-live the creation of the world and all its living beings. Shorn of its religious content, the gayana is still an entertaining, though complicated, tale of magic and adventure. Here is the transcribed text of the gayana, translated from the Bhilali:

I am Malgu gayan.[1] God give me wisdom. From my breast my tongue is moving; on my lap I have a rangai.[2] I will sing Neelsa's song:

The mountains were wild with green. Tigers and bears were roaring. Ranikajal was crying, 'Now what do I do?' She called Ratukamai, 'Devur! Devur![3] Our mountain has changed. What should we do?' Said Ratukamai, 'We should call our mother's priest.' The priest came. He told them to go to the maal and get the singer Malgu.

So who went? Ratukamai did. Went and caught a brown horse. With white reins and spurs of bronze. And a Gujarati saddle. The horse started flying. Ratukamai hit him with a whip of gold. And took the path to Malapur. He reached and called, 'Dada!Dada!'[4] Malgu gayan sleeps for twelve years, snores for thirteen. He awoke with a start, 'Dada, what brings you here?' Said Ratukamai, 'Our task is big. Our mountain is changing. Tigers and bears are

---

[1] The name is derived from *maal*, literally 'shelf', used to describe a piece of flat land in the mountains. A *gayan* is a singer.

[2] A sandglass-shaped drum.

[3] Husband's younger brother.

[4] Literally, elder brother. A term of respect.

roaring. So I have come to take you.' Replied Malgu gayan, 'Go
now, I will come after four–five days.'

Where did Malgu gayan go? To the house of Halva the hutar.[5]
But wood was hard to find. So Malgu made a letter with the dirt
of his chest. Where did he send the letter? To the house of Relu
kabadi.[6] The letter fell on Relu kabadi's chest. He picked it up
and started reading. 'There is a famine of wood. So I have given
you this letter, Relu kabadi. Go.'

'Yes, I shall go', and Relu kabadi got ready. Took some broken
bits of iron. And went to the house of Lahya luhar.[7] An axe of
twelve man,[8] a hammer of thirteen man, a chisel of six man —
all these he got forged. One and a quarter — the price of these
— he gave. Then he went home. Took a tumbda[9] of twelve man,
and water in the tumbda. To his waist he tied food. And took to
the stony path.

Two daughters had Relu kabadi; Revlia and Devlia by name.
They followed him. 'Daughters! Don't come! You must not come!
I go into the big mountains.' They did not listen. Kept going after
him. Relu kabadi went to Vije mountain. Looked all over Vije
mountain; didn't find wood. Dulye mountain, Andaryo mountain, Neelyu mountain, Bhooryo mountain, Janjryo mountain,
Hadyo maal, Dulye maal — 'All mountains have I roamed but I
didn't find teakwood.' Hungry and thirsty, Relu kabadi took the
cloth off his head, spread it on the ground and sat down. 'Tari
mani chudu,[10] I didn't find wood', he started crying.

Then he thought. Picked up some leaves and started worshipping. Prayed and sat. Looked up and saw a Sola teak tree so tall
that it touched the clouds in the heavens! It was growing in the
black clay beside the river. Relu kabadi laughed and smiled.
Brought down the axe and danced four times. One blow of the
axe to the tree and a stream of blood gushed out. So he thought

---

[5] A *hutar* is a carpenter.

[6] A *kabadi* is a woodcutter.

[7] A *luhar* is an ironsmith.

[8] A measure of weight, roughly equal to 36 kilos.

[9] A *tumbda* is a large dried gourd, used to store water or grain.

[10] Bhilali is unabashedly bawdy. People lavishly pepper their speech with
curses such as this which literally means 'fuck your mother'. I have left them
in so that the flavour of the language is retained, but I have not translated them
in the text because they sound unnaturally emphatic when rendered in English.

yet again. Again picked up some leaves of pahal and started praying. Hit another blow with the axe. Water flowed out. 'Tari mani chudu, the mountain does not listen!' Took some more pahal leaves and prayed. Hit another blow with the axe. A stream of milk flowed out. 'Even now the mountain does not listen!' Picked up some more leaves and started praying. Another blow with the axe. The teak tree turned black. 'Now I have it!' Cut five times with the axe. And raised his head to see where the tree would fall.

While Relu kabadi looked up at the tree, his daughters went and hid under the tumbda. The teak tree fell on the tumbda. The tumbda burst and the water flowed out. It became a stream and flowed out. Flowing with the water went Revlia and Devlia to Ambarkhant.[11] Relu kabadi looked towards the tumbda. 'Tari mani chudu! My daughters who were here were washed away.' Ran after them. Went to the sagda bamboo grove and dug there in search of his daughters. Went to the janjra bamboo grove and looked there. Went to the thorny bamboo grove and looked there. Went to the jinta bamboo grove and looked there. Looked everywhere but didn't find his daughters. Ambarkhant was barren; she started caring for the girls, cleaning up after them. She gave them names. The first one she named Ganga, Reva;[12] the second she named Jamna, Tapti,[13] Vijali.

'I didn't find my daughters; I should carry on with the work entrusted to me.' Relu kabadi started breaking branches off the teak tree. First he broke off ganjyo dhol,[14] after that rangai, then heeramangli, then raivajo, then phoolvajo, then dulki, then tutdio, then pipario — all instruments he broke off. Broke them and started carving them. 'All my instruments are done; who shall I give them to? The first letter to come was Malgu gayan's. To him shall I give first the rangai, to Malgu gayan. This worthless drum, it shall be the kumbi's.[15] The ganjyo drum shall be the bazaaria's,

---

[11] Present-day Amarkantak in Shahdol district in Madhya Pradesh, the source of the Narmada.

[12] Ganga and Reva are the two names by which Narmada is referred to in the story.

[13] River Tapti flows south of Narmada, almost parallel to her for a great deal of her length, through the Khandesh region of Maharashtra.

[14] These are names of musical instruments.

[15] Kumbi, dhed and chamar are names of different castes while bazaaria is

the mandal the dhed's. The small pipario shall be the chamar's; the raivajo and the phoolvajo shall go to those who sweep in the markets.' All the instruments he gave away. His wage he took — one and a quarter rupees.

Malgu gayan took the rangai. 'Five days are done; the time of my going has come.' His wife was Mala rani. She made him sit. Heated some water, washed his head and bathed him. Folded him a dhoti of new cloth; dressed him in a bright shirt; tied him an orange turban. 'My going to the mountains has come. Ranikajal has called me; I go to her.' Through deserted lands and peopled lands went Malgu gayan and reached the mountain. Reached Ranikajal who came forward with water to greet him. 'You are really needed, dada; you have reached, it is good', she started unfolding the golden cot. 'O queen! I won't sit there. My place is on the ground', and he sat leaning against the wall. On his lap he had the rangai. 'O queen! Give me a cup of wine.' She served him a cup of wine. Then Malgu gayan prayed. The tigers and bears turned back their ears in alarm. He put a rangai on his lap and started playing. The tigers and bears scattered. The rangai started playing; its sound echoed in the underworld. The sound went to the heavens. And the music reached the ears of Relu kabadi's daughters Revlia and Devlia who started singing. Then Ambarkhant changed their names. One she called Ganga, the other Tapti. Duda hamad[16] had asked for Ganga's hand in marriage, not Tapti's. He gave a coconut to Ganga. The time was set for five days later. Ganga prepared herself and set off to meet the sea.

Then the sisters separated. Said Tapti, 'I shall go by way of Gujarat; you go by Nimar's way.' Ganga came to Rajghat.[17] Being very pious, she said, 'I shall give it a name', and she called it Rajghat. From there she cascaded to Khalghat.[18] After giving it a name she came to Peepalghat. From there came to Onkar maharaj.[19] Onkar maharaj, with pooja things in his hands —

---

used to refer to outsiders, people of the market or city whose caste status is not known. Particular musical instruments are traditionally associated with particular social groups.

[16] The king of the white sea.

[17] This and the places mentioned subsequently are all on the banks of the Narmada.

[18] Khalghat actually lies upstream of Rajghat.

[19] A mendicant.

coconuts and incense — said, 'Mother! Mother! I want to come with you.' Said Ganga, 'No, my son, do not come with me; I am going to Duda hamad's.' When Onkar maharaj would not listen, she said, 'There will be a temple built here for you. As Ambarkhant is our mother, your food and drink will be provided for.'

From there she went forth, cascading to Badwani. From there to Dharamrai. Stopping everywhere made her late. And her sister Tapti reached the sea first and married him. Then a red fish started swimming upstream. From Dharamrai Ganga had come to Kakrana. Said Chauriyo naik,[20] 'Mother! Mother! I am very hungry. Give me something.' And he took his fishing rod and sat. 'My son, I have a lot of creatures in my belly. Sell them and kill them and eat.'

Then the fish met her at Chiloda, 'Bhabhi! Bhabhi![21] What has taken you so long? Tapti has already reached the sea.' 'How did she get ahead of me?' Ganga had a bundle of chivle[22] with her. Because Tapti tied the marriage knot before her, she was enraged. In her rage, she threw away the chivle and turned in to the mountains. Gave names to the banks at Bhitada, 'That one is Bhootvalai, that one Helkaryo, that one Dhudli.' From there she flowed on to Jalhindi. Below that she named Peeparghat. From there she flowed to Gujarat and then into the sea. There she saw that Duda hamad had really wed Tapti. 'What can I do with them?' and she jumped to the bottom. From there she went to Helke ocean, from there to Jelke ocean, from there to Lakhiye ocean, from there to Neelye ocean.

After naming all these oceans, she came to the Relia sand and asked, 'This is sand, barren land. What should I grow here?' So she went to God's garden. What seed did she bring? Thorny white brinjal.[23] Brought the seed and planted it. And to tend it she kept Vasda and Jasda as servants. Day by day the plant grew bigger. Over nine khand[24] grew its nine branches and nine leaves. It was filled with life. And it bore flowers. Their scent went to Kavar

---

[20] Naiks are a low caste people, who live by catching and selling fish. A community of them lives in the village of Kakrana.

[21] Sister-in-law.

[22] A kind of leafy vegetable found by the riverbed.

[23] In a slightly different version of the story, Ambarkhant gives the seed to Narmada as she embarks on her journey.

[24] A unit of area.

land. A brown dog in Kavar land. Said he, 'Tari mani chudu! Where has this scent come from?' That brown dog, he rubbed himself in ashes and became a black wasp. He put on wings of gold and flew away. Came to the heavens. His nose found the scent. Took the path of the oceans.

The nine flowers were flowering away. The black wasp went and fertilized the flowers. After fertilizing them he came away. Then that thorny white brinjal, it started growing fruit. Nine days and nine months, its other flowers stayed barren; nine flowers fruited. First flower fruited; it burst open and from it was born Veelubai. Again a fruit burst — from it was born Jatubai. Another fruit burst — from it was born Katubai. Another fruit burst — Buribai. Another fruit burst — disorders of the stomach came out. Another fruit burst — fever was born. Another fruit burst — epilepsy was born. Another fruit burst — all the poxes came out. Another fruit burst — sores and boils were born.

When Veelubai was small, she crawled. Day by day, she grew and started walking on her feet. Day by day she grew and became very wise. 'I didn't know a mother's womb; I shall make a name. Not a small name; I shall make a big name.' She rubbed some dirt off her chest and made a daughter Bhena kuthar.[25] Saw a dream of making the world.[26] Clay, the son of Bhuinraya — the king of the underworld — lived below the ground. So she started teaching Bhena kuthar. 'Beti,[27] go. You must go to Bhuinraya's house. I have to make the world.' Bhena kuthar went to Bhuinraya's house.

Bhuinraya sleeps for twelve years, snores for thirteen. Bhuinraya was asleep. Bhena kuthar went and stole a lump of clay and brought it to her mother. 'Beti, this won't be enough; go again.' Later, Bhuinraya woke up. 'Tari mani chudu! Who stole my clay?' He took the hair from his forehead, made a noose with it and laid a trap. Bhena kuthar went and stole some more clay. As she came away, she fell into the trap. Bhuinraya went and caught her. 'Beti, beti, why do you steal my clay?' 'Don't take my life; I only labour for someone else.' 'Beti, henceforth don't come this way.' Said Bhena kuthar, 'This work of mine is very important. If you don't

[25] Wasp.

[26] The word used here, kal, refers to Creation, all of the physical world.

[27] Literally, 'daughter', a general term of address for younger women.

give me clay, how will I fulfil my task?' 'Take it then, beti; there is nothing to be done. Take half a lump.'

Bhena kuthar brought the lump to her mother, 'Take it, mother, but from today I won't go to get clay.' 'Why, beti, why won't you go?' asked Veelubai. 'Because Bhuinraya woke up.' Veelubai called her sister Buribai and said, 'Let us start shaping the world.' They prepared the hoe and the spade and fetched water. Started to turn the golden potter's wheel. Turned it to the left and the right. Made the world and held it on the palm of her hand. 'Tari mani chudu, I have shaped the world, but on what should I place it?' Veelubai started crying. From the underworld, the cobra stood up and raised his hood, 'Mother, put your world on my head.' The world was kept and the cobra's hood started trembling. 'I can't hold it up,' said the cobra.

The cow then offered, 'Mother, put your world on my head.' Her horns also bent backwards. 'Mother, you take care of your world; I can't do it.' Veelubai cried even more, 'Who will hold my world?' Then she made nine pillars of wood that covered nine khand and placed the world on it. But they also started shaking and said, 'Mother, take back your world.' Veelubai cried all the more, 'Tari mani chudu, what do I do?' Then a fish, Ragal masa, arched his body into a ring and said, 'Mother, keep your world here.' He took Veelubai's world and said, 'Mother, I will carry your world, but make any creature and place it on my tail.' So Veelubai rubbed some dirt off her chest and made a crab. 'Take this, son, when your tail grows too long and starts coming into your eyes, this crab will keep it trimmed.'

'Now I have a place for my world, but my world is barren. To what creatures should I give shape?' Veelubai started wondering. She made some trees and planted them. 'Beautiful looks my world! I have made all the world, but from where do I make living creatures?' she thought. Took some clay and started shaping creatures. Shaped some lizards. Made tigers and bears. Made snakes. Made men and women. All kinds of creatures she shaped.

'I have shaped everything but my creatures don't have life', Veelubai said and started thinking. As she thought, she rubbed the dirt off her chest. With the dirt from her chest she made a letter. Made the letter and sent it to God's house. God was sleeping. God sleeps for twelve years, snores for thirteen. The letter fell on his chest and he woke up. Started reading the letter. 'I am

a great God; Your name is Veelubai. You have made such big
creatures, tell me the names of your mother and your father and
I will give life to your creatures.' Wrote the letter and sent it to
Veelubai's house.

Veelubai picked up the letter and started reading. Went into
deep thought, 'Tari mani chudu, my mother has a name, but I
don't know my father.' For Veelubai was born of the thorny white
brinjal. She wrote another letter and sent it to God's house, 'I
made the world's creatures but today I don't have a father; I have
a mother but I did not feed at her breast.' Then God wrote a letter
and sent it to her, 'Give me your creatures. A mother is all that
you have; you don't have a father. Since you have made the world,
you have made a name for yourself too.' Veelubai sent a letter
back, 'I will send you my clay creatures. What will you do with
them?' God said, 'They have to have life put into them and blood
put into them.' So Veelubai gave God her creatures of clay.

Now God didn't really know how to put life into creatures.
God had an aunt Banglatrani — his father's sister — who lived
in the twelfth underworld. He sent her a letter and she picked it
up and started reading. Her daughters Lekharia and Zukharia[28]
were studying. Their mother told them, 'Beti, God is the supreme
king. And he has sent us a letter asking how life and blood are to
be put into these creatures.' Banglatrani's daughters said, 'Mother,
we know how to put life into creatures. But God ought to give
us something in exchange.' They sent a letter to God's house. God
replied, 'Take anything you want in return for the life of these
creatures.'

Lekharia and Zukharia said, 'Tari mani chudu, we just wrote
the letter as a joke. But now God has taken us seriously.' They
thought and then wrote to God, 'We know about life, but we
know only half of it, not all.' God replied, 'Beti, even that is a lot.
Somehow or the other, we have to put life in the world.' Lekharia
and Zukharia went around looking for life. There is an ocean —
Vanthar — where the wind blows and makes the ocean speak.
Lekharia and Zukharia set about stealing life from the ocean.
When the wind blew for the ocean to speak, they quickly shut it

---

[28] In the song of the Rathvas, there are the 'female deities Lakhari and Jokhari
who "write and weigh", that is to say, keep account of all the activities on this
earth . . . These deities are supposed to "write" the destiny of each person and
keep a record of his activities' (Jain 1984: 36).

in a box. Then they went to God's house and asked, 'Where are your creatures of clay?' 'They are in the Relia sand. Veelubai shaped them,' said God. Then Lekharia and Zukharia went to the Relia sand and released the wind into the creatures who then came to life. They told God, 'We have put life into them. Now it is up to you. Putting blood into them is your concern.' And God agreed to put blood into the creatures.

Now God had a garden with all kinds of plants in it. Part of his garden was for us humans — this was the garden of juvar. Then God gave breasts to the juvar. Men fed from the breasts and blood flowed into their veins. That is why if we do not eat juvar our blood dries up. For livestock there was a garden of jinjvi[29] grass and God gave it breasts. Livestock also came to have blood. For lizards there was a garden of brown-flowered bengaliphool; God gave it breasts too and blood came into the lizards. But the blood wasn't red; it was brown. For the snakes there was the ningal tree with poisonous flowers of blue and yellow. God gave it breasts too and blood flowed into them. To put blood into the tiger, God made a garden of the ocean — half of milk and half of ghee. That is why the tiger has such a flexible body and can glide anywhere. By feeding in God's garden, all the creatures came to have blood.

'All the world I have now made; all the creatures I have now made, but my world is dark. How can I light it?' and Veelubai fell into thought. There was a Ravut, she sent him a letter, 'How do I light up this world? You must go and take the form of Bhuria gosain.'[30] Then Ravut broke off his own hand and made it into a coconut. Broke off his kneecap to make the hollow body of the rantha.[31] With the skin of his palms he made the stretched skin of the rantha. With his nails he made the bridge of the rantha; he broke his little finger and made pegs for his rantha. With the hair of his head he made strings for the rantha. Took the rantha and started walking. Through deserted lands and peopled lands went Bhuria gosain to Ranubai and Panubai's house.

Ranubai and Panubai were barren from birth; they had no child. 'O maharaj, where have you come from?' 'I have come from roaming the world. Whatever alms people give me along the way,

[29] Dichanthium annulatum.
[30] A gosain is a mendicant.
[31] A musical instrument carried by mendicants.

I have eaten and now I am here.' 'Maharaj, we are barren from
birth. We don't have a son. You won't get any alms here.'
'Ranubai, Panubai, you don't have a son so I shall do some magical
healing.' 'Maharaj, if you give us a son, we have a house of gold
bricks, we will give you even that.' 'No beti, I don't ask for
anything. But you must give me your first child.' 'Yes, maharaj,
we are ready.'

Then Bhuria gosain made a thread. With the thread he tied up
Ranubai and Panubai. 'I will come after nine days and nine
months,' he said and went away. After nine days and nine months
Ranubai and Panubai bore children. Ranubai's first-born was
Surimal, the sun; Panubai's first was Chand, the moon. After that
were born the morning star, then the pole star, then Sirius, after
that Venus. Then the maharaj had to go to them. Veelubai sent
him, saying 'Ravut, don't sit along the way. You must go straight
to Ranubai and Panubai.' Through deserted lands and peopled
lands went Ravut. In a garden was a banyan tree. He went and
sat in its shade.

In that banyan lived a spirit. He stole Ravut's rantha. And hid
in a hollow in the trunk of the tree. The maharaj woke up and
started looking for the rantha. Looked in all the branches and all
the leaves but the spirit lay hidden in the trunk. 'Whoever it be,
man or beast, return my rantha; I am going on Veelubai's work.'
'Look, maharaj, I shall give it back but you must play me a tune.'
So Ravut took the rantha back and started playing a tune. The
spirit started dancing. 'Maharaj, you are going, but give me something.' 'You are a spirit. From now on, you will be invisible to
the human species. This I grant you.' That is why we can't see
spirits now.

Ravut reached Ranubai and Panubai's house. Seeing him coming, they hid their children in seven hundred cells below the
ground. Then they told the maharaj, 'You lied to us. We didn't
get any sons.' Said Ravut, 'No, Ranubai and Panubai, don't lie.
I know you have children.' They quarrelled and fought. 'Very
well, if you won't give them to me, I shall look for your sons
myself.' He started playing the rantha. Surimal's steps were turned.
He came out of his hiding place. Chand's steps were turned. He
came out of his hiding place. One by one they all came out.
'Father, we will come with you.' Said Ranubai and Panubai,
'Maharaj, we will give you whatever you want. Sons, don't go

away.' 'Mother, we were promised. We have to go.' And they took to the path.

Through deserted lands and peopled lands they went. 'Father, go on. We go to bathe in the ocean. We will follow you later.' That is why the sun and the moon rise in the ocean. First the moon, Chand, came out during the day. That day it was so bright and hot that you could cook roti on the tiles. Then Veelubai said, 'Surimal, you are the sun. You will come out during the day and Chand will come out during the night.'

A Glossary of Hindi and Bhilali Words

Andolan	Social movement
Bazaaria	Adivasi term for non-adivasis
Begaar	Forced, unpaid labour
Bhagat	Hinduized adivasi
Bhajan	Hindu devotional song
Bhangjadya	Broker of village quarrels
Bidi	Tobacco rolled in a leaf and smoked
Budva	Shaman
Chabutra	Platform
Chaumasa	Rainy season
Daakan	Witch
Dahia	Shifting cultivation
Gayana	Song of creation
Ghee	Clarified butter
Haat	Weekly market
Huru	Liquor
Jagir	Estate conferred by a king on his chieftain
Jiji	Sister
Juvar	Sorghum
Kushta	Loincloth
Laah	Labour sharing
Laahtia	Middleman
Lathi	Stick

Maal	Platform, overhead shelf
Maanta	Vow
Motla	Big, important
Naik	Hereditary tribal chief
Nakedar	Forest or revenue official
Nevad	'Encroached' cultivated land
Paat	Irrigation channel
Panchayat	a) Traditional village council
	b) Village-level elected government
Pannia	*Roti* cooked between leaves
Parikrama	Circumambulation
Patel	Village headman
Patidar	A caste of dominant landowners in the plains
Patvari	Official maintaining village land and revenue records
Phalya	Hamlet
Pooja	Worship
Pujara	Priest
Raabdi	Fermented maize gruel
Rantha	Violin-like musical instrument
Sangath	Union
Sankalp	Resolve
Satyagraha	Gandhian passive resistance
Subedar	Non-commissioned officer in the army
Tehsil	Sub-district in modern India
Thakur	Ruler of minor estate
Thakurat	Minor estate
Vaataad	Spokesman
Vania	Trader-moneylender
Vikas	Development

Bibliography

Alvares, Claude and Ramesh Billorey. 1987. 'Damming the Narmada: The Politics Behind the Destruction' in *The Ecologist*. Vol. 17: No. 2.

Amin, Shahid. 1988. 'Gandhi as Mahatma' in R. Guha and G.C. Spivak (eds) *Selected Subaltern Studies*. Oxford University Press: New York.

Anderson, Benedict. 1983. *Imagined Communities: Reflections on the Origins and Spread of Nationalism*. Verso: London.

Aurora, G.S. 1972. *Tribe-Caste-Class Encounters: Some Aspects of Folk-Urban Relations in Alirajpur Tehsil*. Administrative Staff College: Hyderabad.

Bahro, Rudolf. 1982. *Socialism and Survival*. Heretic Books: London.

Bailey, F.G. 1960. *Tribe, Caste and Nation*. Manchester University Press: Manchester.

Banwari. 1992. *Pancavati: Indian Approach to Environment*. Translated from Hindi by Asha Vohra. Shri Vinayak Publications: Delhi.

Baviskar, Amita. 1991. 'Narmada "Sangharsh Yatra": State's Response and its Consequences' in *Economic and Political Weekly*. Vol. 26: Nos. 9–10.

Benton, Ted. 1989. 'Marxism and Natural Limits: An Ecological Critique and Reconstruction' in *New Left Review*. No. 178.

Berger, S. 1979. 'Politics and Antipolitics in Western Europe' in *Daedalus*. Winter.

Béteille, Andre. 1986. 'The Concept of Tribe with Special Reference to India' in *European Journal of Sociology*. XXVII.

Bhatia, Bela. 1993. 'Forced Evictions of Tribal Oustees due to the Sardar Sarovar Project in Five Submerging Villages of Gujarat'. Report submitted to the Gujarat High Court. Mimeo.

Bhattacharyya, P.K. 1977. *Historical Geography of Madhya Pradesh from Early Records*. Motilal Banarsidass: Delhi.

Bose, N.K. 1971. *Tribal Life in India*. National Book Trust: Delhi.

Bourdieu, Pierre. 1977. *Outline of a Theory of Practice*. Cambridge University Press: Cambridge.

Breman, Jan. 1985. *Of Peasants, Migrants and Paupers: Rural Labour Circulation and Capitalist Production in West India.* Oxford University Press: Delhi.

Byres, T.J. 1981. 'The New Technology, Class Formation and Class Action in the Indian Countryside' in *Journal of Peasant Studies.* Vol. 8: No. 4.

Canak, William. 1989. 'Debt, Austerity, and Latin America in the New International Division of Labour' in Canak (ed.) *Lost Promises.*

Chambers, Robert, N.C. Saxena and Tushaar Shah. 1989. *To the Hands of the Poor: Water and Trees.* Intermediate Technology Publications: London.

Champion, Sir H.G. and S.K. Seth. 1968. *A Revised Survey of the Forest Types of India.* Government of India Press: Nasik.

Chandra, Sudhir. 1987. 'To My Successor . . . ' in *Times of India: Sunday Review.* 8 November 1987.

Chatterjee, P. 1989. 'Caste and Subaltern Consciousness' in R. Guha (ed.) *Subaltern Studies VI: Writings on South Asian History and Society.* Oxford University Press: Delhi.

Chatwin, Bruce. 1987. *The Songlines.* Pantheon: New York.

Clark, John P. 1989. 'Marx's Inorganic Body' in *Environmental Ethics.* Vol. 11: No. 3.

Clifford, James. 1983. 'On Ethnographic Authority' in *Representations.* Vol. 1: No. 2.

———. 1990. 'Introduction: Partial Truths' in J. Clifford and G.E. Marcus (eds) *Writing Culture: The Poetics and Politics of Ethnography.* Oxford University Press [1986].

CSE (Centre for Science and Environment). 1985. *The State of India's Environment 1984–85: The Second Citizens' Report.* New Delhi.

———. 1991. *Floods, Flood Plains and Environmental Myths.* New Delhi.

Deliege, Robert. 1985. *The Bhils of Western India: Some Empirical and Theoretical Issues in Anthropology in India.* National: Delhi.

Dewey, Clive. 1978. '*Patwari* and *Chaukidar*: Subordinate Officials and the Reliability of India's Agricultural Statistics' in C. Dewey and A.G. Hopkins (eds) *The Imperial Impact: Studies in the Economic History of Africa and India.* The Athlone Press: London.

DG (Dhar Gazetteer). 1984. *Gazetteer of India: Madhya Pradesh: Dhar.* Director, Gazetteers Madhya Pradesh: Bhopal.

Dharmadhikary, Shripad. 1991. 'The Narmada Dams Controversy: Issues for Energy Policy' in *Urja.* Vol. 29: No. 5.

Dumont, Louis. 1970. *Homo Hierarchicus: An Essay on the Caste System.* University of Chicago Press: Chicago.

Epstein, S.J.M. 1988. *The Earthy Soil: Bombay Peasants and the Indian Nationalist Movement 1919–1947.* Oxford University Press: Delhi.

Evans, Peter B. and John D. Stephens. 1988. 'Development and the World Economy' in N.J. Smelser (ed.) *The Handbook of Sociology.* Sage: California.

Freire, Paulo. 1970. *Pedagogy of the Oppressed.* Seabury Press: New York.

Gadgil, Madhav and Ramachandra Guha. 1992. *This Fissured Land: An Ecological History of India.* Oxford University Press: Delhi.

Gandhi, M.K. 1951. *Towards Non-Violent Socialism.* Edited by B. Kumarappa. Navajivan Publishing House: Ahmedabad.

Geertz, Clifford. 1988. *Works and Lives: The Anthropologist as Author.* Stanford University Press: Stanford.

George, Susan. 1988. *A Fate Worse Than Debt.* Penguin: London.

Ghurye, G.S. 1963. *The Scheduled Tribes.* Popular Prakashan: Bombay.

GOI (Government of India). 1978. *Provisions in the Constitution of India for Scheduled Tribes.* Ministry of Home Affairs: New Delhi.

——. 1981. *Census of India 1981. Series 11 – Madhya Pradesh.* Part XIII-A. Village and Town Directory. Jhabua District: District Census Handbook.

Goldsmith, Edward and Nicholas Hildyard. 1984. *The Social and Environmental Effects of Large Dams: A Report to the European Ecological Action Group.* Wadebridge Ecological Centre: Camelford.

Gore, Al. 1992. *Earth in the Balance: Forging a New Common Purpose.* Viva Books: New Delhi.

Gouldner, Alvin. 1979. *The Future of Intellectuals and the Rise of the New Class: A Frame of Reference, Theses, Conjectures, Arguments, and an Historical Perspective on the Role of Intellectuals and the Intelligentsia in the International Class Contest of the Modern Era.* Oxford University Press: New York.

Guha, Ramachandra. 1988. 'Ideological Trends in Indian Environmentalism' in *Economic and Political Weekly.* Vol. 23: No. 49.

——. 1989a. *The Unquiet Woods: Ecological Change and Peasant Resistance in the Himalaya.* University of California Press: Berkeley.

——. 1989b. 'Radical American Environmentalism and Wilderness Preservation: A Third World Critique' in *Environmental Ethics.* Vol. 11: No. 1.

——. 1992. 'Prehistory of Indian Environmentalism: Intellectual Traditions' in *Economic and Political Weekly.* Vol. 27: Nos. 1 & 2.

Guha, Ramachandra and Madhav Gadgil. 1989. 'State Forestry and Social Conflict in British India' in *Past and Present: A Journal of Historical Studies.* No. 123.

Guha, Ranajit. 1988. 'The Prose of Counterinsurgency' in R. Guha and G.C. Spivak (eds) *Selected Subaltern Studies.* Oxford University Press: New York.

Guru, Shambhu Dayal. 1983. *Madhya Pradesh ki Nadiyan* [Rivers of Madhya Pradesh]. J.K. Publishing House: Bhopal.

Habermas, Jurgen. 1981. 'New Social Movements' in *Telos.* No. 49.

Hardiman, D. 1987a. 'The Bhils and Shahukars of Eastern Gujarat' in R. Guha (ed.) *Subaltern Studies V: Writings on South Asian History and Society.* Oxford University Press: Delhi.

———. 1987b. *The Coming of the Devi: Adivasi Assertion in Western India.* Oxford University Press: Delhi.

Herring, Ronald. 1991. *Politics of Nature: Interests, Commons Dilemmas and the State.* Paper presented at Conference on Common Property, Collective Action and Ecology, Centre for Ecological Sciences, Indian Institute of Sciences, Bangalore. August 1991.

Hobsbawm, E.J. 1959. *Primitive Rebels: Studies in the Archaic Forms of Social Movement in the 19th and 20th Centuries.* W.W. Norton: New York.

Jain, Jyotindra. 1984. *Painted Myths of Creation: Art and Ritual of an Indian Tribe.* Lalit Kala Akademi: New Delhi.

Joshi, Vidyut. 1991. *Rehabilitation — A Promise to Keep: A Case of SSP.* The Tax Publications: Ahmedabad.

Kalpavriksh. 1988. *The Narmada Valley Project: A Critique.* New Delhi.

Kalpavriksh and Hindu College Nature Club. 1985. 'The Narmada Valley Project: Development or Destruction?' in *The Ecologist.* Vol. 15: No. 5/6.

KG (Khandesh Gazetteer). 1880. *Gazetteer of the Bombay Presidency.* Vol. XII. Government Central Press: Bombay.

Kohli, Atul. 1987. *The State and Poverty in India: The Politics of Reform.* Cambridge University Press: Cambridge.

Kothari, Rajni. 1988. *State Against Democracy: In Search of Humane Governance.* Ajanta Publishers: Delhi.

Kulkarni, D.S. 1983. 'The Bhil Movement in Dhulia District (1972-74)' in K.S. Singh (ed.) *Tribal Movements in India.* Vol. 2. Manohar: Delhi.

LCHR (Lawyers Committee for Human Rights). 1993. *Unacceptable Means: India's Sardar Sarovar Project and Violations of Human Rights: October 1992 through February 1993.* New York.

Leach, E.R. (ed.) 1960. *Aspects of Caste in South India, Ceylon and North-West Pakistan.* Cambridge.

Lincoln, Yvonna S. and Egon G. Guba. 1985. *Naturalistic Inquiry.* Sage Publications: Newbury Park.

Lipsky, Michael. 1968. 'Protest as a Political Resource' in *American Political Science Review.* Vol. 62: No. 4.

Luard, C.E. 1908a. *Central India Gazetteer Series: Western States (Malwa).* Vol. v – Part A. Text. British India Press: Bombay.

———. 1908b. *Central India Gazetteer Series: Western States (Malwa).* Vol. v – Part B. Tables. British India Press: Bombay.

———. 1912. *Minor States (Thakurats) Gazetteer.* Standard Press: Allahabad.

Ludden, D. 1984. 'Productive Power in Agriculture: A Survey of Work on the Local History of British India' in M. Desai, S.H. Rudolph and A. Rudra (eds) *Agrarian Power and Agricultural Productivity in South Asia.* University of California Press: Berkeley.

Lukes, Steven. 1975. 'Political Ritual and Social Integration' in *Sociology.* Vol. 9: No. 2.

Manohar, D. n.d. *Aamu Aanhi Vaagha Pille* [We Are Tiger Cubs: A History of the Khandesh Bhils]. Translated from Marathi into Hindi by Jayashree Bhalerao Bhatnagar.

Marcus, George E. 1990. 'Contemporary Problems of Ethnography in the Modern World System' in J. Clifford and G.E. Marcus (eds) *Writing Culture: The Poetics and Politics of Ethnography.* Oxford University Press [1986].

MARG (Multiple Action Research Group). 1986. *Sardar Sarovar Oustees in Madhya Pradesh: What Do They Know? (I) Alirajpur.* New Delhi.

Marris, Peter. 1987. *Meaning and Action: Community Planning and Conceptions of Change.* Routledge and Kegan Paul: London.

Marx, Karl and Friedrich Engels. 1852. 'The Eighteenth Brumaire of Louis Bonaparte' in *Collected Works.* Progress: Moscow.

Mayer, Adrian C. 1960. *Caste and Kinship in Central India.* University of California Press: Berkeley.

Mohanty, Gopinath. 1987. *Paraja* [Translated from Oriya into English by Bikram K. Das]. Oxford University Press.

Moraes, Dom. 1983. *Answered by Flutes: Reflections from Madhya Pradesh.* Asia Publishing House: Bombay.

Morse, Bradford and Thomas Berger. 1992. *Sardar Sarovar: The Report of the Independent Review.* Resource Futures International: Ottawa.

Nag, Moni. 1984. 'Fertility Differential in Kerala and West Bengal: Equity-Fertility Hypothesis as Explanation' in *Economic and Political Weekly.* Vol. 19: No. 1.

Nandy, Ashis. 1987. *Traditions, Tyranny and Utopias: Essays in the Politics of Awareness.* Oxford University Press: Delhi.

Nath, Y.V.S. 1960. *Bhils of Ratanmal: An Analysis of the Social Structure of a Western Indian Community.* The M.S. University of Baroda: Baroda.

NBA (*Narmada Bachao Andolan*). 1991. *Towards Sustainable and Just Development: The Peoples' Struggle in the Narmada Valley.* Mimeo.

NBA Delhi (*Narmada Bachao Andolan*: Delhi). *Narmada: A Campaign Newsletter.* Nos. 1–10. Delhi.

NCA (Narmada Control Authority). 1991. *Submergence of Villages in Gujarat, Maharashtra and Madhya Pradesh with the Construction of SSP.* Indore.

NCAER (National Council of Applied Economic Research). 1963. *Socio-Economic Conditions of Primitive Tribes in Madhya Pradesh.* NCAER: New Delhi.

O'Connor, James. 1988. 'Capitalism, Nature, Socialism: A Theoretical Introduction' in *Capitalism, Nature, Socialism.* No. 1.

Offe, Claus. 1985. 'New Social Movements: Challenging the Boundaries of Institutional Politics' in *Social Research.* Vol. 52: No. 4.

Omvedt, Gail. 1987. 'India's Green Movements' in *Race and Class.* Vol. 28: No. 4.

———. 1992. 'Ecofeminism in Action: Healing India with Women's Power' in *Guardian.* 25 March.

Pangare, Ganesh and Vasudha. 1992. *From Poverty to Plenty: The Story of Ralegan Siddhi.* INTACH: New Delhi.

Parajuli, Pramod. 1991. 'Power and Knowledge in Development Discourse: New Social Movements and the State in India' in *International Social Science Journal.* No. 127.

Paranjpye, Vijay. 1990. *High Dams on the Narmada: A Holistic Analysis of the River Valley Projects.* Indian National Trust for Art and Cultural Heritage: New Delhi.

Pocock, D. 1972. *Kanbi and Patidar: A Study of the Patidar Community of Gujarat.* Clarendon Press: Oxford.

Prasad, Ashok and Harish Dhawan. 1982. *A Sanctuary for Birds Only.* Kalpavriksh: New Delhi.

PUCL (People's Union for Civil Liberties, Madhya Pradesh). 1990. *Vikas ka Aatank* [The Terror of Development].

Ram, Rahul N. 1993. *Muddy Waters: A Critical Assessment of the Benefits of the Sardar Sarovar Project.* Kalpavriksh: New Delhi.

Redclift, Michael. 1987. *Sustainable Development: Exploring the Contradictions.* Methuen: London.

Redford, Kent H. 1991. 'The Ecologically Noble Savage' in *Cultural Survival Quarterly*. Vol. 15: No. 1.

Roseberry, William. 1989. *Anthropologies and Histories: Essays in Culture, History and Political Economy*. Rutgers University Press: New Brunswick.

Russell R.V. and Hiralal. 1916. *Tribes and Castes of the Central Provinces of India* [1975 ed.] Cosmo Publications: Delhi.

Sachidanandan, P. 1988. 'National Perspective for Irrigation Development in India' in J.S. Kanwar (ed.) *Water Management – The Key to Developing Agriculture*. Indian National Science Academy: New Delhi.

Sangvi, Sanjay and Alok Agrawal. 1991. 'State Repression in Madhya Pradesh: Target — Popular Movements' in *Economic and Political Weekly*. Vol. 26: No. 47.

Sarkar, Sumit. 1980. 'Primitive Rebellion and Modern Nationalism: A Note on Forest Satyagraha in the Non-Cooperation and Civil Disobedience Movements' in K.N. Panikkar (ed.) *National and Left Movements in India*. Vikas: New Delhi.

Sayer, Derek. 1987. *The Violence of Abstraction: The Analytic Foundations of Historical Materialism*. Basil Blackwell: Oxford.

Scott, James C. 1976. *The Moral Economy of the Peasant: Rebellion and Subsistence in Southeast Asia*. Yale University Press: New Haven.

——. 1985. *Weapons of the Weak: Everyday Forms of Peasant Resistance*. Yale University Press: New Haven.

——. 1990. *Domination and the Arts of Resistance: Hidden Transcripts*. Yale University Press: New Haven.

Sengupta, Nirmal. 1993. *User-Friendly Irrigation Designs*. Sage: New Delhi.

Shah, Ghanshyam and Arjun Patel. 1993. *Tribal Movements in Western India: A Review of Literature*. Centre for Social Studies: Surat.

Sharma, B.D. 1990. *Report of the Commissioner for Scheduled Castes and Scheduled Tribes*. Twenty-ninth Report 1988–89.

Sharma, Suresh. 1992. 'The Vanquished Tribal World of Shifting Cultivation' in A. Bhalla and P.J. Bumke (eds) *Images of Rural India in the 20th Century*. Sterling Publishers: New Delhi.

Sheth, Pravin. 1991. 'The Politics of Ecofundamentalism' in *All About Narmada*. Directorate of Information, Government of Gujarat: Gandhinagar.

Shiva, Vandana and J. Bandyopadhyay. 1990. 'Asia's Forests, Asia's Cultures' in S. Head and R. Heinzman (eds) *Lessons of the Rainforest*. Sierra Club Books: San Francisco.

Singh, Chhatrapati. 1986. *Common Property and Common Poverty: India's Forests, Forest Dwellers and the Law.* Delhi.

Spivak, Gayatri Chakravorty. 1988. 'Subaltern Studies: Deconstructing Historiography' in Ranajit Guha and G.C. Spivak (eds) *Selected Subaltern Studies.* Oxford University Press: New York.

Stacey, Judith. 1988. 'Can There be a Feminist Ethnography?' in *Women's Studies International Forum.* Vol. 11: No. 1.

Stiglmayr, E. 1970. *The Barela-Bhilalas and their Songs of Creation.* Acta Ethnologica et Linguistica: Vienna.

Taylor, Charles. 1985. 'Rationality' in *Philosophy and the Human Sciences: Philosophical Papers 2.* Cambridge University Press: Cambridge.

Taylor, Kenneth I. 1990. 'Why Supernatural Eels Matter' in S. Head and R. Heinzman (eds) *Lessons of the Rainforest.* Sierra Club Books: San Francisco.

Thompson, E.P. 1975. *Whigs and Hunters: The Origins of the Black Act.* Pantheon: New York.

Tilly, Charles. 1985. 'War Making and State Making as Organized Crime' in P.B. Evans, D. Rueschemeyer and T. Skocpol (eds) *Bringing the State Back In.* Cambridge University Press: New York.

TISS (Tata Institute of Social Sciences). 1993. 'Sardar Sarovar Project: Review of Resettlement and Rehabilitation in Maharashtra' in *Economic and Political Weekly.* Vol. 28: No. 34.

UNDP (United Nations Development Programme). 1992. *Human Development Report 1992.* Oxford University Press: New York.

UNDP (United Nations Development Programme). 1993. *Human Development Report 1993.* Oxford University Press: New York.

Varma, S.C. 1978. *The Bhil Kills.* Kunj Publishing House: Delhi.

Vatsyayan, Kapila. 1992. 'Ecology and Indian Myth' in G. Sen (ed.) *Indigenous Vision: Peoples of India, Attitudes to the Environment.* Sage: New Delhi.

Vohra, B.B. 1980. *A Policy for Land and Water.* Sardar Patel Memorial Lectures: Department of Environment, Government of India.

Watts, Michael. n.d. *Sustainability and Struggles over Nature: Political Ecology or Ecological Marxism?* Mimeo.

Webster, Andrew. 1984. *Introduction to the Sociology of Development.* Humanities Press: New Jersey.

Winch, Peter. 1964. 'Understanding a Primitive Society' in *American Philosophical Quarterly.* No. 1.

WNG (West Nimar Gazetteer). 1970. *Gazetteer of India: Madhya Pradesh: West Nimar.* District Gazetteers Department, Madhya Pradesh: Bhopal.

Wolf, Eric R. 1982. *Europe and the People Without History*. University of California Press: Berkeley.

Unpublished Sources

NAI (National Archives of India), Foreign Department.
NAI (National Archives of India), Bhopawar Political Agency.

Index